VALORIZATION OF LIQUID WASTE AS ALTERNATIVE FUEL

AF482557

Pritam Dey

LIST OF TABLES

LIST OF FIGURES

LIST OF SYMBOLS AND ABBREVIATIONS

AD	Anderson-Darling
ADP	Abiotic deletion potential
AETP	Aquatic eco-toxicity potential
ANOVA	Analysis of variance
AP	Acidification potential
ASTM	American Standard for Testing and Material
BBD	Box-Behnken Design
BP	Brake power
BSFC	Brake specific fuel consumption
BTE	Brake thermal efficiency
CCI	Calculated cetane index
CI	Compression ignition
CO	Carbon monoxide
CO_2	Carbon dioxide
CV	Calorific value
D100	pure diesel
D-WVO	Diesel - waste vegetable oil blend
D-WVO-BD	Diesel - waste vegetable oil biodiesel blend
D-WVO_0.20	0.80 fraction diesel + 0.20 fraction WVO
D-WVO_0.35	0.65 fraction diesel + 0.35 fraction WVO
D-WVO_0.50	0.50 fraction diesel + 0.50 fraction WVO
D-WVO-BD_0.20	0.80 fraction diesel + 0.20 fraction WVO-BD
D-WVO-BD_0.35	0.65 fraction diesel + 0.35 fraction WVO-BD
D-WVO-BD_0.50	0.50 fraction diesel + 0.50 fraction WVO-BD
EP	Eutrophication potential
FC	Fuel consumption
FT-IR	Fourier transform infrared spectroscopy
GC-MS	Gas chromatography-Mass spectroscopy
GWP	Global warming potential
HC	Unburnt hydrocarbon
HTP	Human toxicity potential
ID	Ignition delay
ICE	Internal Combustion Engine
IOCL	Indian Oil Corporation Limited

IS	Indian standards
K100	pure kerosene
K-WVO_0.20	0.80 fraction kerosene + 0.20 fraction WVO
K-WVO_0.35	0.65 fraction kerosene + 0.35 fraction WVO
K-WVO_0.50	0.50 fraction kerosene + 0.50 fraction WVO
ME	Mechanical efficiency
MUFA	Monounsaturated fatty acid
NO_x	Oxide of nitrogen
NHR_{max}	maximum net heat release rate
ODP	Ozone depletion potential
OX	Oxidability
p	probability function
P_{max}	maximum cylinder pressure
PUFA	polyunsaturated fatty acid
PEI	Potential environmental impact
R^2	Coefficient of correlation
RS	Response surface
RSM	Response surface methodology
TE	Thermal efficiency
TETP	Terrestrial eco-toxicity potential
VE	Volumetric efficiency
WFO	Waste frying oil
WVO	Waste vegetable oil
WVO_0H	Straight vegetable oil
WVO_24H	Waste vegetable oil, thermally exposed for 24 hours
WVO_48H	Waste vegetable oil, thermally exposed for 48 hours
WVO_72H	Waste vegetable oil, thermally exposed for 72 hours
WVO-BD	Waste vegetable oil biodiesel
WVO-BD_24h	Waste vegetable oil biodiesel, prepared from WVO with 24h of thermal exposure time
WVO-BD_48h	Waste vegetable oil biodiesel, prepared from WVO with 48h of thermal exposure time
WVO-BD_72h	Waste vegetable oil biodiesel, prepared from WVO with 72h of thermal exposure time

CONTENTS

Chapter 1

INTRODUCTION

1.1 Background

The demand for liquid fuel in transportation, agriculture, and power generation sectors has seen a steep rise in the last decade. International Energy Agency has reported that the reason for the sudden rise in fuel demand is driven by industrialization, modernization, and rapid economic development [1]. Moreover, according to the annual report of British Petroleum, published in 2019, the worldwide fuel oil consumption has increased from 3.93 billion tons to 4.33 billion tons in the last 10 years [2]. The surge in fossil fuel demand is particularly prominent in developing countries like India. It has been predicted that by 2030, the world would require 1.5 times the energy which it consumes today, and India and China are believed to consume more than 40% of that energy [3].

Diesel is the fuel of choice for industrial, power generation, logistics, and agricultural sectors, due to its ability to produce more energy at high compression ratios [4]. The rise in diesel demand has resulted in the rise in diesel prices and price instability throughout the world. Also, to meet this growing demand the fossil reserves are depleted rapidly [5]. Hence, proving a renewable diesel fuel alternative is the need of the hour. Moreover, the refining process for diesel production and the combustion of diesel in engines are causing irreparable harm to the environment. Thus, there has been a growing environmental concern over diesel combustion, which has pressed for research towards renewable and alternative diesel substitutes [6]. From 2010 to 2015, India has recorded a 64.15% increase in the number of on-road vehicles [7]. As a result, there has been more than a 70% increase in the consumption of diesel fuel in the last few years [8]. A rise in diesel demand and depletion of fossil fuel reserves are the primary requisites towards the search for renewable diesel fuel substitutes which can significantly reduce diesel consumption.

However, complete independence from fossil fuels is unrealized to date and complete replacement for diesel is also not reported. In this regard, the blending of renewable fuel with fossil fuel (especially diesel) has gained significance [9, 10]. Vegetable oils and their biodiesels are often blended with kerosene, another fossil fuel, in place of diesel. Such blends are investigated for their properties and engine performance, and emission characteristics by fueling diesel engines. [11, 12]. A variety of edible and non-edible oils were considered and studied to provide diesel fuel alternatives [13, 14]. However, utilization of vegetable oils for fuel generation offsets the requirement to feed the ever-growing population – leading to an ongoing fuel versus food dispute [15]. Thus, the current focus has shifted towards identifying and utilizing those resources which are outside the fuel versus food nexus. In this regard, waste vegetable oil is the most widely regarded solution to all the aforementioned problems [16, 17].

Waste vegetable oil (WVO) is a term used invariably throughout the article to denote waste cooking oil, waste frying oil, or yellow grease. The primary producers of WVO are packaged food manufactures, food processing units, hotels, restaurants, and fast food outlets [18, 19]. The production of WVO has been increasing each year throughout the world. It is estimated that around 24.3 million tons of cooking oil are used in India per year, out of which 1.3 million tons of WVO can be collected as feedstock for biodiesel production [20]. Moreover, the United Nations Food and Agricultural Organization has estimated the potential for recovery of 8.0 million tons of WVO from 23.0 million tons of vegetable oil used in the food processing sector [21].

The generation of such huge quantities of WVO poses threats to the environment and human health. Such oils cannot be used any further in food preparation, due to oxidation and the presence of suspended or dissolved food particles [22]. As a waste, WVO cannot be discharged directly into the environment or wastewater because it is reported to cause clogging of the collection pipes and sanitary sewer overflow. Additionally, disposal of WVO in landfills and waterways leads to soil and water pollutions and also disrupts the aquatic ecosystems [23].

Nevertheless, as useless and wasteful as it sounds, WVO can be a potential feedstock for WVO-biodiesel (WVO-BD). As an additional benefit, WVO is inexpensive, and all the costs incurred would be in collecting and treating only [24]. Recycling WVO as an alternative fuel for diesel engines also presents a promising avenue to reduce dependency on depleting fossil oil reserves [25, 26]. The key to the successful utilization of WVO is establishing a reliable and continuous supply of WVO.

WVO is one of the most prudent energy sources because of its restrictions for food use, low cost, wider availability, and pertinence for fuel production [27]. The application of WVO for alternative fuel production addresses environmental issues concerning the handling and safe disposal of WVO [28] but also generates clean and renewable energy when used as a fuel source [29]. Conversion of WVO into WVO-BD by transesterification is a known method to valorize WVO. Several reports are available on the utilization of WVO as feedstock for biodiesel production. Numerous studies have reported the efficacy of WVO-BD as an alternative fuel fraction in diesel to run compression ignition (CI) engines [16]. Countless reports are available in the literature on straight vegetable oil blends with diesel to fuel CI engines [13, 14]. Additionally, vegetable oil blends with kerosene were also investigated in CI engines [11, 12]. However, the direct and unmodified blending of WVO with fossil fuels is unexplored to date.

Thus, the thesis provides several studies to valorize WVO as an alternative fuel fraction in CI engines. Firstly, the wide-accepted derivative form of WVO, WVO-BD, is optimized to be blended with different fossil fuels to attain density and calorific value similar to diesel for fueling a CI engine. Then a comparative study between diesel-WVO-BD blends and WVO-diesel blends is presented to demonstrate the potential of WVO as an alternative fuel fraction in diesel. An optimization study of diesel-WVO blends to model fuel consumption and thermal efficiency in a CI engine is presented. Following it, an elaborative study on the performance and emission characteristics of a CI engine fueled with separate WVO blends with both diesel and kerosene is presented. In the final study, a comparative analysis among various diesel-WVO and kerosene-WVO blends is performed to ascertain the WVO blend most suitable in terms of engine combustion, performance, emission characteristics, engine running cost, and potential environmental impacts associated with the blends.

1.2 Research gaps and scopes

The utilization of WVO in the production of biodiesel, known as WVO-biodiesel (WVO-BD) through the transesterification process has been the focus of many researchers throughout the globe. The composition of WVO varies from region to region. But, in a locality, the composition of WVO remains the same. Thus, utilization of locally available WVO can be a good research scope. In this regard, proper characterization of WVO in terms of fatty acid composition and physical properties is a prerequisite for any study.

WVO-BD is known to be blended with diesel to fuel CI engines. Vegetable oil biodiesels are reportedly blended with other fossil fuels like kerosene and gasoline. Hence, the blending of WVO-BD with various fuel oils and the changes in their physical properties have not been studied yet. So, it serves as another research gap. Optimizing important physical properties of WVO-BD blends in different fossil fuel oils to define a diesel substitute is scope for novel research.

Also, it has been proven that vegetable oils can be used to produce blends for diesel engines. Such blends have performed well in existing diesel engines, without any modification, and have resulted in nominal wear of the engine. Thus, there is a possibility that if and when the properties of WVO are properly controlled, the same can be used as an inexpensive alternative blend fuel for fueling an unmodified CI engine. No study in the open literature deals with the utilization of WVO (without preheating or additive addition) to be used as an alternative fraction in blend fuel for diesel engines. So the same is another research gap in this field.

Additionally, a comparative analysis of WVO-BD and WVO, in terms of fuel economy, emissions, and engine performance is nowhere to be found in the open literature. This comparison is very much necessary to answer a pertinent question – Is it necessary for WVO to be transesterified before blending it with diesel? Or can chemically unmodified WVO as a blend fraction in diesel can be used to operate low-duty CI engines? Thus, it is also an important research scope to investigate.

Finally, a research scope exists in blending WVO and kerosene to run a CI engine. The scientific logic involved in this hypothesis is that if vegetable oils in kerosene blends have been studied to run CI engines, why can't kerosene-WVO blends do the same? In this regard, a comprehensive comparison between diesel-WVO and kerosene-WVO blends in terms of engine combustion, performance, and emissions is a logical approach. Also, comparative assessment based on engine operating cost and potential environmental impacts linked with the various WVO blends is an appropriate research opportunity.

1.3 Thesis objective

Thus the main objectives of the thesis are:

1. To develop a waste vegetable oil biodiesel-based diesel substitute fuel by response surface optimization of density and calorific value.

   ➔ WVO-BD mixed in various fossil fuels at various proportions will elucidate varying physical properties. Thus, to valorize WVO-BD as an alternative fraction in blend fuels to replace diesel, an optimization of the two most important fuel properties is of immense significance.

2. To comparatively analyze waste vegetable oil and waste vegetable oil-biodiesel in diesel as an alternative fuel for a CI engine.

   ➔ Can WVO be utilized as a diesel fuel fraction without chemical modifications? This question can be answered when WVO-diesel blends are compared against WVO-BD-diesel blends for engine performance, emission, and operating cost.

3. To optimize fuel consumption and thermal efficiency of a CI engine fueled with waste vegetable oil-diesel blends.

   ➔ The appropriate diesel blend proportion of WVO, of varying thermal exposure time, is necessary to be optimized to achieve improved engine performance.

4. To valorize waste vegetable oil for utilization as blend fuel in a CI engine.

   ➔ As WVO is established as an alternative fuel fraction in blend fuels, a full-scale study of diesel blends containing WVO of varying thermal exposure time as an alternative fuel for CI engines is a pertinent research objective. Also, blending WVO with kerosene and investigating the engine performance and emission outcomes of such blends is a unique research objective.

5. To comparatively analyze waste vegetable oil blends with diesel and kerosene.

   ➜ Finally, a comparative assessment of WVO-diesel and WVO-kerosene blends is very much necessary to choose the most sustainable WVO blend fuel which has the highest potential to substitute diesel.

1.4 Thesis organization

The thesis is divided into eight (08) chapters. Chapter 1 introduces the brief background, the knowledge gap, the research questions that arise in formulating the thesis, the main objectives of the thesis, and lastly the organization of the thesis is listed.

Chapter 2 discusses briefly the literature relevant to this work and identifies the knowledge gaps so that the work can be justified. Also, the experimental methodologies for achieving the objectives were formulated from the state of art.

Chapter 3 follows the development of a response surface model for predicting the density and calorific value of the WVO-BD blend with various fossil fuels. Box-Behnken design of experimental array is selected for carrying out the modeling of density, from which calorific value is derived. An optimization study is also presented and a comparison of the model-optimized WVO-BD blend with diesel is shown.

In chapter 4, a comparison is drawn between various diesel blends of WVO-BD and WVO in terms of fuel properties, engine performance, and emission characteristics. The economics of utilizing WVO-diesel blends in place of the popular WVO-BD-diesel blends is presented at the end of the chapter.

Chapter 5 presents a modeling and optimization study to develop a response surface model for fuel consumption and thermal efficiency of compression ignition engine fueled with diesel blends with different fractions of WVO at different engine load conditions. The model-optimized WVO-diesel blend for low fuel consumption and high thermal efficiency is compared with diesel.

Chapter 6 discusses the valorization of WVO as an alternative fraction in fossil fuels. Firstly, the engine performance and emission characteristics of WVO, of three thermal treatment durations, after blending in diesel are presented. Followed by the engine performance and emissions of WVO-kerosene blends are reported.

In chapter 7, a comparative study of WVO blends in diesel and kerosene is presented. The comparison was based on the variations in the fuel characteristics, engine combustion, performance, and emission characteristics of WVO-diesel and WVO-kerosene blends. In addition to that, the engine running cost and potential environmental impacts associated with D-WVO and K-WVO blends are also considered for a comprehensive analysis of WVO blends.

In conclusion, chapter 8 provides a summary of the dissertation and highlights; the contributions made from this work. Also, the possible future scopes of the research work are suggested.

1.5 References

[1] International Energy Agency, (IEA) (2019), "World Energy Outlook 2017: China, December 2019" Accessed on Oct 26, 2020. https://www.iea.org/reports/world-energy-outlook-2017-china

[2] Dudley, B. (2018), "BP statistical review of world energy" BP Statistical Review, London, UK. (Accessed Aug 6, 2018)

[3] Uguz, G., Atabani, A. E., Mohammed, M. N., Shobana, S., Uguz, S., Kumar, G., Al-Mahtaseb, A. H. (2019), "Fuel stability of biodiesel from waste cooking oil: A comparative evaluation with various antioxidants using FT-IR and DSC techniques", Biocatal Agri Biotechnol, v. 21, pp. 101283.

[4] Reşitoğlu, İ. A., Altinişik, K., Keskin, A. (2015), "The pollutant emissions from diesel-engine vehicles and exhaust after-treatment systems", Clean Technol Environ Policy, v. 17, pp. 15–27.

[5] Tamilselvan, P., Nallusamy, N., Rajkumar, S. (2017), "A comprehensive review on performance, combustion and emission characteristics of biodiesel fueled diesel engines", Renew Sustain Energy Rev, v. 79, pp. 1134-1159.

[6] Mahmudul, H. M., Hagos, F. Y., Mamat, R., Abdul Adam, A., Ishak, W. F. W., Alenezi, R. (2017), "Production, characterization, and performance of biodiesel as an alternative fuel in diesel engines – A review", Renew Sustain Energy Rev, v. 72, pp. 497-509.

[7] Ministry of Statistics and Program Implementation (MOSPI). 2016. Government of India report. Accessed August 20, 2018. http://mospi.nic.in/sites/default/files/statistical_year_book_india_2015/Table-20.1_0.xlsx.

[8] Petroleum Planning and Analysis Cell. Ministry of Petroleum and Natural Gas, Government of India data (2018). (Accessed August 20, 2018). http://ppac.org.in/View_All_Reports.aspx

[9] Suresh, M., Jawahar, C. P., Richard, A. (2018), "A review on biodiesel production, combustion, performance, and emission characteristics of non-edible oils in variable compression ratio diesel engine using biodiesel and its blends", Renew Sustain Energy Rev, v. 92, pp. 38-49.

[10] Mofijur, M. G. R. M., Rasul, A. M., Hyde, J., Azad, A. K., Mamat, R., Bhuiya, M. M. K. (2016), "Role of biofuel and their binary (diesel–biodiesel) and ternary (ethanol–biodiesel–diesel) blends on internal combustion engines emission reduction", Renew Sustain Energy Rev, v. 53, pp. 265-278.

[11] Labeckas, G., Slavinskas, S. (2015), "Combustion phenomenon, performance and emissions of a diesel engine with aviation turbine JP-8 fuel and rapeseed biodiesel blends", Energy Convers Manag, v. 105, pp. 216-229.

[12] Bayindir, H., Isık, M. K., Argunhan, Z., Yücel, H. L., Aydın, H. (2017) "Combustion, performance, and emissions of a diesel power generator fueled with biodiesel-kerosene and biodiesel-kerosene-diesel blends", Energy, v. 123, pp. 241-251.

[13] Dabi, M., Saha, U. K. (2019), "Application potential of vegetable oils as an alternative to diesel fuels in compression ignition engines: A review", J Energy Instit, v. 92, pp. 1710-1726.

[14] Mat, S. C., Idroas, M. Y., Hamid, M. F., Zainal, Z. A. (2018), Performance and emissions of straight vegetable oils and its blends as a fuel in diesel engine: A review", Renew Sustain Energy Rev, v. 82 pp. 808-823.

[15] Capuano, D., Costa, M., Di Fraia, S., Massarotti, N., Vanoli, L. (2017), "Direct use of waste vegetable oil in internal combustion engines", Renew Sustain Energy Rev, v. 69 pp. 759-770.

[16] Kathirvel, S., Layek, A., Muthuraman, S. (2016), "Exploration of waste cooking oil methyl esters (WCOME) as fuel in compression ignition engines: a critical review", Eng Sci Technol Int J, v. 19, pp. 1018-1026.

[18] Fonseca, J. M., Teleken, J. G., Almeida, V. C., da Silva, C. (2019), "Biodiesel from waste frying oils: Methods of production and purification", Energy Conver Manag, v. 194, pp. 205-218.

[19] Moazeni, F., Chen, Y-C., Zhang, G. (2019), "Enzymatic transesterification for biodiesel production from used cooking oil, A review", J Clean Prod, v. 216, pp. 117-128.

[20] PTI, (2019), "Government launches program for converting used cooking oil into biodiesel in 100 cities", Economic Times, Published Aug 10, 2019.

[21] Food and Agricultural Organization of the United Nations, (2020), "Oilseeds and Oils and Meals - Monthly Price updates", Report of January 126, pp. 1-14. Accessed 07 July 2020.

[22] Refaat, A. A., Attia, N. K., Sibak, H. A., El Sheltawy, S. T., El-Diwani, G. I. (2008), "Production optimization and quality assessment of biodiesel from waste vegetable oil", Int J Environ Sci Technol, v. 5(1), pp.75-82.

[23] Mohammed, A. R., Bandari, C. (2020), "Lab-scale catalytic production of biodiesel from waste cooking oil – a review", Biofuel, v. 11, pp. 409-419.

[24] Senthil Kumar, M., Jaykumar, M. A. (2014), "Comprehensive study on performance, emission and combustion behavior of a compression ignition engine fueled with WCO emulsion as fuel", J Energy Institute, v. 87, pp. 263–271.

[25] Dhanasekaran, R., Krishnamoorthy, V., Rana, D., Saravanan, S., Nagendran, A., Rajesh Kumar, B. (2017), "A sustainable and eco-friendly fueling approach for direct-injection diesel engines using restaurant yellow grease and n-pentanol in blends with diesel fuel", Fuel, v. 193, pp. 419–431.

[26] Krishnamoorthy, V., Dhanasekaran, R., Rana, D., Saravanan, S., Kumar, B. R. (2018), "A comparative assessment of ternary blends of three bio-alcohols with waste cooking oil and diesel for optimum emissions and performance in a CI engine using RSM", Energy Convers Manag, v. 156, pp. 337-357.

[27] Li, L., Ding, Z., Li, K., Xu, J., Liu, F., Liu, S., Yu, S., Xie, C., Ge, X. (2016), "Liquid hydrocarbon fuels from catalytic cracking of waste cooking oils using ultrastable zeolite USY as a catalyst", J Anal Appl Pyrol, v. 117, pp. 268–272.

[28] Abuhabaya, A., Fieldhouse, J. D., Brown, D. R. (2010), "Evaluation of properties and use of waste vegetable oil (WVO), pure vegetable oils and standard diesel as used in a compression ignition engine", In Future Technologies in Computing and Engineering Annual Researchers' Conference (CEARC'10), The University of Huddersfield, pp. 71-76.

[29] Wako, F. M., Reshad, A. S., Bhalerao, M. S., Goud, V. V. (2018), "Catalytic cracking of waste cooking oil for biofuel production using zirconium oxide catalyst", Indus Crop Prod, v. 18, pp. 282-289.

Chapter 2

LITERATURE REVIEW

The literature review primarily deals with the state-of-the-art in utilizing diesel blends with vegetable oils of different origins and biodiesels derived from waste vegetable oil in running unmodified compression ignition (CI) engines. The scientific logic behind presenting such a literature review is to show that vegetable oils as blend fractions in diesel (or in few cases kerosene) are suitable in replacing diesel for running CI engines. Also, the recent trend on utilizing biodiesel, derived from waste vegetable oil (WVO), as an alternative fraction in diesel needs to be looked into before making appropriate hypotheses. The review presents brief discussions on the research findings from recent literature ahead of presenting the review in a tabulated manner. The review is focused on the changes in engine performance and emission characteristics of CI engines when fueled with diesel blends containing varying fractions of vegetable oil or waste vegetable oil biodiesel (WVO-BD), compared with diesel. In addition to that, it should be reiterated that no study on the direct blending of WVO without preheating or additives is present in the literature. This marks the novelty of the thesis. Thus, this chapter presents a systematic review on the following topics:

2.1 Utilization of vegetable oil blends in CI engines

2.2 Utilization of waste vegetable oil biodiesel blends in CI engines

2.3 Utilization of waste vegetable oil in CI engine

2.1 Utilization of vegetable oil blends in CI engines

The utilization of vegetable oils in compression ignition (CI) engines has been widely reported. Vegetable oils are mainly extracted from the seeds and also from the kernel of food crops. Vegetable oil can be divided into two categories, which are edible and non-edible oils. Rapeseed, sunflower, corn, soybean, coconut, peanut, sesame, and palm oil are examples of edible oil, while jatropha, mahua, karanja, linseed, jojoba, poon, rubberseed, and cottonseed oils are examples of non-edible oil. However, the direct use of vegetable oil in CI engines is not advisable due to several limitations. Relatively higher density (0.88-0.96 g/cm^3 at 15 °C), greater viscosity (28-62 mm^2/s at 40 °C), and lower calorific values (35.5-41.7 MJ/kg) than diesel are a few of those limitations. Also, vegetable oil is linked with poor cold flow properties, inferior volatility, and engine knocking than commercial diesel fuel. Also, coking and carbon deposition in fuel lines, on injector nozzle, piston and engine head, cylinder walls, deterioration of lubricating oil are some of the long-term issues which are reportedly observed with vegetable oil.

Researchers have suggested preheating the oil before injection, blending with diesel or other additives, and transesterification (to produce biodiesel) as some of the possible routes to overcome such impediments. Blending vegetable oils with diesel (or other fossil fuel) has become a widely accepted choice when the idea is to avoid increasing the fuel price due to the transesterification process. Several reviews, published recently, have focused only on the blends of various edible and non-edible oils with diesel and their engine performance and emission studies [1-3].

Selected literature on the changes in engine performance and emission characteristics for various kinds of vegetable oil blends in fossil fuel is shown in Table 2.1. The vegetable oil fraction in the blends ranged from as low as 0.10 (for Tamanu oil [4]) to as high as 0.60 (for Cottonseed oil [5]). Apart from Huang et al. (2016) [6] and How et al. (2015) [7], the vegetable oil blends were assessed to run light (3.5-7.2 kW) to medium-duty (12-18 kW) CI engines. While most vegetable oils were blended with diesel, a couple of reports are available for vegetable oil blending in kerosene [8, 9]. In general, a rise in the brake-specific fuel consumption (BSFC) and a fall in the brake thermal efficiency (BTE) with vegetable oil fractions in diesel blends was concluded. However, the opposite was reported for vegetable oil-based kerosene blends, where the kerosene-vegetable oil blends demonstrated higher BTE than diesel.

Table 2.1 Selected literature (2012-2020) on the reported changes in engine performance and emission characteristics of vegetable oil blends.

Veg oil fraction	Engine operating parameters*	Change in performance ^		Change in emission ^			References
		BSFC	BTE	CO	HC	NO$_x$	
0.50 fraction Jatropha oil in diesel	12 kW BP, 4-s, 2-cyl, at full load, 17.5:1 CR, 1500 RPM	↑7.3%	↑4.3%	↑41.7%	----	↑15%	Sidibe et al. 2020 [10]
0.50 fraction Fennel seed oil in diesel	6.3 kW, 4-s, 1-cyl, 18:1 CR, 3000 RPM	----	↓15.6%	↑33.3%	↑16%	↑17.4%	Hazar et al. 2019 [11]
0.40 fraction Nahar seed oil in diesel	3.5 kW, 4-s, 1-cyl, 16.5:1 CR, 210 bar IP, 23° bTDC IT, 1500 RPM	↑14.5%	↓12.4%	↑100%	↑16.7%	↓11.8%	Dash and Lingfa 2018 [12]
0.20 fraction Cottonseed oil in diesel	5.2 kW, 4-s, 1-cyl, at 80% load, 17.5:1 CR, 210 bar IP, 1500 RPM	↑0.8%	↑1.2%	↑50%	↔	↑6.7%	Raj et al. 2018 [13]
0.20 fraction Thumba oil in diesel	3.5 kW, 4-s, 1-cyl, 1:1 CR, 210 bar IP, 23° bTDC IT, 1500 RPM	↑6.1%	↓2.4%	↑16.7%	↓20%	↓9.8%	Jain et al. 2017 [14]

0.20 fraction Rubberseed oil in diesel	4.4 kW, 4-s, 1-cyl, 17.5:1 CR, 220 bar IP, 24° bTDC IT, 1500	↑3.6%	↓15.4%	↓28.7%	↑12.3%	↑14.1%	Karthik et al. 2017 [15]
0.25 fraction of Lemongrass oil in diesel	3.5 kW, 4-s, 1-cyl, at full load, 17.5:1 CR, 210 bar IP, 1500 RPM	↑30.7%	↓12.5%	↑20%	↑26.2%	↑20.7%	Sathiyamo orthi and Sankaranar ayanan 2017 [16]
0.10 fraction Tamanu oil in diesel	3.5 kW, 4-s, 1-cyl, 17:1 CR, 220 bar IP, 1500 RPM	↑8.3%	↓8.3%	↓50%	↓5.2%	↑75%	Yarrapraga da and Krishna 2017 [4]
0.50 fraction Canola oil in kerosene	18 kW 4s, 4 cyl. CI engine at 60% load, 17:1 CR, 1500 RPM	↓2.4%	↑1.7%	↑108%	↑127%	↓76%	Bayindir et al. 2017 [8]
0.40 fraction Cardanol oil in kerosene	3.5 kW 4s, 1 cyl. CI engine at full load and 1500 RPM	↓15.1%	↑9.8%	↑200%	↑6.7%	↓16.7%	Ravindra et al. 2017 [9]
0.50 fraction Pine oil in diesel	55 kW, 4s, 4 cyl. CI engine at 60% load and 1800 RPM	↑3.6%	↓3.7%	↓53.3%	↓6.9%	↑7.9%	Huang et al. 2016 [6]
0.60 fraction Cottonseed oil in diesel	3.5 kW 4s, 1 cyl. CI engine at full load and 1500 RPM	----	↓5.6%	↓10.5%	↑10%	↓1.2%	Martin et al. 2016 [5]
0.40 fraction Mahua oil in diesel	3.7 kW 4s, 1 cyl. CI engine at full load and 1500 RPM	↑5.3%	↓8.0%	↓20%	↓25%	↑3.6%	Sonar et al. 2015 [17]
0.50 fraction Rapeseed oil in diesel	13.7 kW, 4s, 2 cyl. CI engine at full load and 1500 RPM	↑2.8%	↓4.1%	↓33.3%	↑10%	↑5.3%	Qi et al. 2014 [18]
0.50 fraction Karanja oil in diesel	7.2 kW 4s, 4 cyl. CI engine at full load and 1500 RPM	↑40.8%	↓25%	↓94%	↓77%	↑24.5%	Agarwal and Dhar 2013 [19]
0.50 fraction Coconut oil in diesel	38 kW, 4s, 4 cyl. CI engine at full load and 5000 RPM	↑17.9%	----	↓18.2%	↓27.3%	↓6.2%	How et al. 2012 [7]

Note: * - considered at the highest reported engine load condition
^ - change is relative to neat diesel fuel at similar engine operating conditions
↑ - increase in parameter, ↓ - decrease in parameter, ↔ - no change
BP – brake power, bTDC – before top dead center, CI – compression ignition, s – stroke, cyl – cylinder, CR – compression ratio, IP – injection pressure, RPM – rotations per minute.

2.2 Utilization of WVO-BD blends in CI engines

WVO has been hailed as a possible substitute for vegetable oil for the production of biodiesel. Such biodiesel is termed as WVO-biodiesel (WVO-BD). WVO-BD is the methyl ester of the triglycerides or fatty acids present in the oil (WVO). The process of conversion of WVO into WVO-BD is known as transesterification. In this process, the fatty acids in WVO are esterified using alcohol (generally methanol) in the presence of a catalyst and temperature. The resultant products are the methylated esters of fatty acids and glycerol. Two types of catalysts - acid catalysts (e.g. sulfuric acid) and base catalysts (e.g. sodium hydroxide) – are employed for the process.

The utilization of WVO-BD as an alternative fraction for diesel fuel in a CI engine has been a thrust area in energy research for the past decade or so. Numerous reports were published in this regard. A comprehensive review of the engine performance and emission outcomes of WVO-BD was last published in the year 2016 by Kathirvel et al. [2]. After that, no significant review is published to date. Thus, a detailed literature survey on the recent reports on engine performance and emission characteristics of WVO-BD is relevant. Reports on the applications of WVO-BD before 2015 were not considered in the literature survey under this review article. A comprehensive and insightful literature survey report is presented in Table 2.2, where the changes in engine performance and emissions are denoted relative to diesel fuel.

As evident from Table 2.2, the WVO-BD fraction in diesel usually ranged from 0.10 to 0.50. Only a few researchers have utilized WVO-BD fractions higher than 0.50 in CI engines. Apart from Chuah et al. (2015) [47], Wei et al. (2018) [38], and Wu et al. (2019) [30], WVO-BD blends with diesel were tested at light or medium-duty CI engines. WVO-BD fractions in diesel are reported to increase the BSFC and decrease the BTE of the engine, relative to neat diesel fuel. The slightly higher density of WVO-BD blends as compared to diesel resulted in marginally lower volatility of such blends, causing improper atomization through the conventional injector [31]. As a result, high mass fractions of such blends were required to generate the required power, resulting in higher BSFC values of WVO-BD blends, as compared to straight diesel fuel.

Most of the reports on WVO-BD diesel blends in CI engines revealed reductions in both CO and HC emissions. The CO reduction varied from as low as 3.5% [25] to as high as 47.8% [38], relative to diesel. Whereas, the HC emission was reported to have reduced by 4.0 – 41.6 %. Only a handful of reports have suggested increased CO and HC emissions for WVO-BD blends when compared to neat diesel [24, 41]. The outcome of NO_x emission is debatable, as a significant number of contrasting reports are available in the literature. Interestingly, Abed et al. (2018) [32] have reported a 32.5% rise, whereas, Sabariswaran and Selvakumar (2017) [42] have concluded a 48.7% reduction in NO_x emission compared to neat diesel.

Table 2.2 Literature survey (2015-2020) on the reported changes in engine performance and emission characteristics of biodiesel produced from waste vegetable oil.

WVO-BD fraction	Engine operating parameters*	Change in performance ^		Change in emission ^			References
		BSFC	BTE	CO	HC	NO_x	
0.50	11 kW BP, 4-s, 4-cyl, 17:1 CR, 1500 RPM	↑0.6%	↑7.8%	↑50%	↓4%	↓18.6%	Karayaka 2020 [24]
0.50	35.5 kW BP, 4-s, 6-cyl, 16:1 CR, 197.5 bar IP 1500 RPM	-----	↑1.2%	↓25%	-----	↑13.6%	Hojati and Shirneshan 2020 [25]
0.80	9 kW, 4-s, 1-cyl, 17.5:1 CR, 210 bar IP, 1400 RPM	↑18.2%	↓4.2%	↓3.5%	↓23.2%	↑7.3%	Ors 2020 [26]
0.20	3.7 kW BP, 4-s, 1-cyl, 17.5:1 CR, 220 bar IP, 1500 RPM	↑1.1%	↓5%	-----	↑5%	↓5%	Al-Dawody et al. 2019 [27]
0.20	33.5 kW BP, 4-s, 2-cyl, 18.5:1 CR, 500 bar IP, 2250 RPM	----	↓4%	↓15%	↓5.5%	↑12%	Mathew et al. 2019 [28]
0.50	3.5 kW BP, 4-s, 1-cyl, 17.5:1 CR, 200 bar IP, 1500 RPM	↑3.5%	↓1.2%	↓4.8%	↓4.2%	↑2.8%	Singh et al. 2019 [29]
0.10	234 kW BP, 4-s, 6-cyl, 17:1 CR, 240 bar IP, 1500 RPM	↑1.6%	↓0.9%	↓25.2%	↓41.6%	↑10.7%	Wu et al. 2019 [30]
0.30	4.56 kW BP, 4-s, 1-cyl, 18:1 CR, 210 bar IP, 1500 RPM	↑11.5%	↓10.3%	↓13.8%	↓19.2%	↑29.5%	Yesilyurt 2019 [31]
0.30	4 kW BP, 4-s, 1-cyl, 17.5:1 CR, 175 bar IP, 1500 RPM	↑6.7%	↓2%	↓33.3%	↓40%	↑32.5%	Abed et al. 2018 [32]

0.15	4.2 kW BP, 4-s, 1-cyl, 17.5:1 CR, 190 bar IP, 1500 RPM	↑7%	↓2.1%	↓13.6%	↓5.4%	↑1%	Borugadda et al. 2018 [33]
0.50	46.4 kW BP, 4-s, 4-cyl, 1500 RPM	↑11%	-----	↓47.8%	-----	↑27%	Garcia-Martin et al. 2018 [34]
0.20	11 kW BP, 4-s, 1-cyl, 18:1 CR, 235 bar IP, 2000 RPM	↑5.2%	↓1.6%	↓28.5%	↓27.3%	↑7.2%	Kumar and Gakkhar 2018 [35]
0.60	3 kW BP, 4-s, 1-cyl, 17.5:1 CR, 1500 RPM	↑4.7%	↑4%	-----	-----	-----	Ray and Prakash 2018 [36]
0.50	24 kW BP, 4-s, 1-cyl, 22.5:1 CR, 1800 RPM	↑4.4%	↓0.5%	↓35.2%	↑163.6%	↑7.6%	Ulusoy et al. 2018 [37]
0.50	134 kW BP, 6-s, 1-cyl, 17:1 CR, 210 bar IP, 1500 RPM	↑5.1%	↓1.5%	-----	-----	↓6.3%	Wei et al. 2018 [38]
0.70	136 kW BP, 6-s, 1-cyl, 17:1 CR, 210 bar IP, 1500 RPM	↑8.3%	-----	-----	-----	↓8.2%	Geng et al. 2017 [39]
0.50	25 kW BP, 4-s, 4-cyl, 16:1 CR, 2000 RPM	↑26.7%	↓16%	-----	-----	-----	Meisami et al. 2017 [40]
0.40	4.7 kW BP, 4-s, 1-cyl, 17.5:1 CR, 210 bar IP, 1500 RPM	↑40%	↓23%	↑355%	↓41.5%	↑32%	Prabu et al. 2017 [41]
0.20	4.4 kW BP, 4-s, 1-cyl, 17.5:1 CR, 1500 RPM	↑45.5%	↓28%	↓25%	↓40%	↓48.7%	Sabariswaran and Selvakumar 2017 [42]
0.50	0.73 MPa BMEP, 4-s, 4-cyl, 17.5:1 CR, 1500 RPM	↑3.3%	-----	↑7.4%	↑31%	-----	Wei et al. 2017 [43]
0.20	3.7 kW BP, 4-s, 1-cyl, 16:1 CR, 1500 RPM	↑4%	↓1.6%	↑5%	-----	↑7.7%	Yildizhan et al. 2017 [44]
0.50	5.8 kW BP, 4-s, 1-cyl, 1500 RPM	↑8%	↓2%	↓50%	↓20%	↓6%	Attia and Hassaneen 2016 [45]
0.40	3.7 kW BP, 4-s, 1-cyl, 1500 RPM	↑16.7%	↓9.7%	↔	↓6.3%	↑4%	Kotebavi et al. 2016 [46]
0.30	300 kW BP, 4-s, 6-cyl, 17:1 CR, 2000 RPM	↑8.3%	↓8.4%	↓26.3%	-----	↑19%	Chuah et al. 2015 [47]

Note: * - considered at the highest reported engine load condition
^ - change is with respect to neat diesel fuel at similar engine operating conditions
↑ - increase in parameter, ↓ - decrease in parameter, ↔ - no change
BP – brake power, BMEP – Brake mean effective pressure, CI – compression ignition, s – stroke, cyl – cylinder, CR – compression ratio, IP – injection pressure, RPM – rotations per minute.

2.3 WVO utilization as alternative fuel fraction

WVO utilization as an alternative fuel fraction is a much-neglected niche of alternative fuel research. A handful of articles are available in the literature which showed direct utilization (with various additives) of WVO as the alternative fraction of diesel blend. Table 2.3 enlists the literature on this topic. It is noteworthy that the lack of motivation and associated research on direct WVO utilization is the prime reason for selecting WVO as the alternative fraction for blend fuels. A rise in brake-specific fuel consumption (BSFC) and a fall in brake thermal efficiency (BTE) was concurrently reported in the literature. In terms of emissions, a rise in unburnt hydrocarbons (HC) was persistent. However, there were mixed outcomes for carbon monoxide (CO) and oxides of nitrogen (NO_x).

WVO after a few pre-treatment steps has been found to behave like straight vegetable oils. Such pre-treatments include filtration using a fine sieve, gravity settling of suspended particles and repeated washing with water for the removal of water-soluble food molecules. Drying at or above $105 \pm 5°C$ is also performed to remove any residual moisture in the WVO. Additionally, several advanced pretreatment methods can be used, such as steam injection, neutralization, vacuum evaporation, and vacuum filtration [48].

Only a handful of reports are available on the utilization of WVO as diesel blends. But, in all these studies, WVO was used either after preheating or with additives [48]. WVO was reportedly pre-heated at 55, 65, or 75 °C before injection [49, 50]. Few researchers have performed studies on ternary blends containing WVO, diesel, and alcohol [51, 52]. The alcohol was blended to lower the flash point, density, and viscosity and to increase the volatility of the blend to achieve a faster pre-mixing and a wider pre-mixed burning phase [51]. Krishnamoorthy et al. (2018) [53] have performed an RSM-based optimization study on the effects of three different alcohols – n-propanol, n-butanol, and n-pentanol - on the engine performance and emission characteristics of diesel-WVO blends. Senthilkumar and Jaykumar (2014) [54] have studied the performance and emission characteristics of a CI engine of emulsified palm oil-based WVO in diesel blends emulsified with water, ethanol, and surfactants.

The literature on the direct utilization of WVO (with preheating or additives) as an alternative blend fraction for diesel is summarized in Table 2.3. As evident from the table, a 4.1 - 7.2% rise in brake specific fuel consumption (BSFC) and a 3.8 - 10.7% fall in brake thermal efficiency (BTE) were observed for diesel-WVO blends as compared to neat diesel fuel. Neat WVOs, preheated at 55 - 75 °C, was reported to result in an 11.8 – 20.0% increase in brake specific energy consumption (BSEC) coupled with around 10.7% decrease in BTE, relative to diesel. An increase in carbon monoxide (CO) emission was reported for diesel-WVO blends mixed with either alcohol or as an emulsion in water, by many researchers [51, 52, 54]. However, around 6.5% decrease in CO was reported by D'Alessandro et al. (2016) [55], while no change in CO reduction was concluded by Hribernik and Kegl (2019) [56]. Likewise, in the case of hydrocarbon (HC) emissions on the contrary to the general upward trend, Kalam et al. (2011) [57] have reported a 27.3 – 36.4% reduction in HC emissions. Whereas, the emission of oxides of nitrogen (NO_x) for the WVO blends was reported to elucidate mixed outcomes – higher NO_x emissions in some studies, while lower NO_x emissions in others, as compared to diesel.

Table 2.3 Literature survey on direct WVO utilization in CI engine as diesel fraction (upon preheating, as emulsion, or with additives).

WVO blend type	Engine operating parameters	Changes in engine performance and emissions*	References
0.50 fraction WVO-diesel blend with n-pentanol	5.2 kW, 0.661 L, 4-s, 1-cyl, DI CI at 210 bar IP, 23° bTDC IT, 16.5:1 CR, 1500 RPM	7.1% ↑ in BSFC, 3.8% ↓ in BTE, 50% ↑ in CO, 41.2% ↑ in HC, 6.7% ↓ in NOx	Dhanasekaran et al. 2017 [51]
0.50 fraction WVO-diesel blend with n-pentanol	4.4 kW, 0.661 L, 4-s, 1-cyl, DI CI at 210 bar IP, 23° bTDC IT, 17.5:1 CR, 1500 RPM	5% ↑ in BSFC, 5% ↓ in BTE, 50% ↑ in CO, 41.2% ↑ in HC, 7.2% ↓ in NOx	Ravikumar and Saravanan 2017 [52]
Neat Sunflower WVO preheated at 65 °C before injection	33 kW, 3.11 L, 4-s, 1-cyl, DI turbo-charged CI at 210 bar IP, 23° bTDC IT, 17.5:1 CR, 1500 RPM	9.8% ↑ in BSEC, 3.2% ↓ in CO, 28.1% ↑ in NOx	D'Alessandro et al. 2016 [55]
Neat Palm WVO preheated at 65 °C	33 kW, 3.11 L, 4-s, 1-cyl, DI turbo-charged CI at 210 bar IP,	11.8% ↑ in BSEC, 6.5% ↓ in CO, 22.4% ↑ in NOx	D'Alessandro et al. 2016

before injection	23° bTDC IT, 17.5:1 CR, 1500 RPM		[55]
Neat WVO emulsions in diesel	3.7 kW, 0.630 L, 4-s, 1-cyl, DI CI at 200 bar IP, 27° bTDC IT, 16:1 CR, 1500 RPM	14.1% ↓ in BTE, 60% ↑ in CO, 25% ↑ in HC, and 24.4% ↓ in NOx	Senthilkumar and Jaykumar 2014 [54]
0.05 fraction Palm WVO-diesel blend	53.6 kW, 2.4 L, 4-s, 4-cyl, IDI CI at 22.3:1 CR, 3000 RPM	1.2% ↓ in BP, 27.3% ↓ in HC, and 2.1% ↑ in NOx	Kalam et al. 2011 [57]
0.05 fraction Coconut WVO-diesel blend	53.6 kW, 2.4 L, 4-s, 4-cyl, IDI CI at 22.3:1 CR, 3000 RPM	0.7% ↓ in BP, 36.4% ↓ in HC, and 0.8% ↓ in NOx	Kalam et al. 2011 [57]
0.75 fraction WVO-diesel blend preheated before injection	40 kW, 1.590 L, 4-s, 4-cyl, IDI CI at 190 bar IP, 27° bTDC IT, 23.5:1 CR, 1500 RPM	No change in CO, 13.6% ↑ in HC, and 125% ↑ in NOx	Hribernik and Kegl 2009 [56]
Neat WVO preheated above 75 °C before injection	3.7 kW, 4-s, 1-cyl, DI CI at 190 bar IP, 27° bTDC IT, 1500 RPM	20% ↑ in BSEC, 10.7% ↓ in BTE, 163.6% ↑ in CO, and 19.6% ↑ in NOx	Pugazhvadivu and Jeyachandran 2005 [49]
Neat WVO preheated above 55 °C before injection	4.0 kW, 4-s, 1-cyl, DI CI at 3600 RPM	4.5% ↑ in BSFC, 8.8% ↓ in BP, 2% ↓ in BTE, 8.7% ↑ in CO, and 13.5% ↑ in NOx	Bari et al. 2002 [50]

Note: L - liter; s – stroke; cyl – cylinder; CR – compression ignition; DI – direct injection; RPM – rotations per minute; IDI – indirect injection; IP – injection pressure; IT – injection timing; bTDC = before top dead center; EGR – exhaust gas recirculation; BSFC – brake specific fuel consumption; BSEC – brake specific energy consumption; BP – brake power; BTE – brake thermal efficiency; CO – carbon monoxide; HC – hydrocarbon; NO_x – oxides of nitrogen; NR – not reported; ↑ - increase; ↓ - decrease; * - relative to neat diesel fuel at the highest engine load used in the study.

From the literature survey, it has been proven that vegetable oils can be used to produce blends for diesel engines and such blends have performed well in existing unmodified diesel engines. Thus, there is a possibility that WVO, if properly controlled, can be used as an alternative fraction of the blend. No study in the open literature deals with the utilization of WVO (without preheating or additive addition) to be used as an alternative fraction in blend fuel for diesel engines. Finally, a comparative analysis of WVO-BD and WVO, in terms of fuel economy, emissions, and economics is nowhere to be found in the open literature. Thus, it is also a research scope to investigate. Also, blending WVO with kerosene is still unexplored, which provides another scope for research. Thus, it can be inferred that the valorization and utilization of locally available WVO can be a good research scope.

Thus, following a systematic and exhaustive literature review the following hypotheses are formulated:

(i) A model to predict the density and calorific value of WVO-BD blends, when it is mixed with fossil fuels of varying densities, can be developed. And an optimized diesel fuel substitute based on density and calorific value can be presented.

(ii) There is no need for WVO to be transesterified into WVO-BD before being used in diesel blends. Unmodified WVO in diesel may be used to operate light-duty CI engines without any difficulties.

(iii) If straight vegetable oils fractions up to 0.50 in diesel were used to operate low-duty (or low power) CI engines, then unmodified or unadulterated WVO blends in diesel can do the same.

(iv) Kerosene can be a better fossil fuel oil for preparing blends with WVO for utilization in low-duty CI engines. Such blends are more sustainable than diesel-WVO blends.

2.4 References

[1] No, S. Y. (2017). "Application of straight vegetable oil from triglyceride-based biomass to IC engines–A review". Renew Sustain Energy Rev, v. 69, pp. 80-97.

[2] Mat, S. C., Idroas, M. Y., Hamid, M. F., Zainal, Z. A. (2018), "Performance and emissions of straight vegetable oils and its blends as a fuel in diesel engine: A review", Renew Sustain Energy Rev, v. 82 pp. 808-823.

[3] Dabi, M., Saha, U. K. (2019), "Application potential of vegetable oils as alternative to diesel fuels in compression ignition engines: A review", J. Energy Institute, v. 92, pp. 1710-1726.

[4] Yarrapragada, K. S. S. R., Krishna, B. B. (2017), "Impact of tamanu oil-diesel blend on combustion, performance and emissions of diesel engine and its prediction methodology", J. Brazilian Soc. Mech. Sci. Eng., v. 39, pp. 1797–1811.

[5] Martin, M. L. J., Geo, V. E., Nagalingam, B. (2016), "Effect of fuel inlet temperature on cottonseed oil-diesel mixture composition and performance in a DI diesel engine", J. Energy Institute, v. 90, pp. 563-573.

[6] Huang, H., Teng, W., Liu, O., Zhou, C., Wang, Q., Wang, X. (2016), "Combustion, performance and emission characteristics of a diesel engine under low-temperature combustion of pine oil–diesel blends", Energy Convers. Manag., v. 128, pp. 317-326.

[7] How, H. G., Teoh, Y. H., Masjuki, H. H., Kalam, M. A. (2012), "Impact of coconut oil blends on particulate-phase PAHs and regulated emissions from a light duty diesel engine", Energy, v. 48, pp. 500-509.

[8] Bayindir, H., Isık, M. Z., Argunhan, Z., Yücel, H. L., Aydın, H. (2017), "Evaluation of combustion, performance and emission indicators of canola oil-kerosene blends in a power generator diesel engine", Applied Thermal Eng., v. 114, pp. 234–44.

[9] Ravindra, M., Aruna, Vardhan, H. (2017), "Investigation on the performance of a variable compression ratio engine operated with raw cardanol kerosene blends", Biofuels, pp. 1-7.

[10] Sidibe, S., Blin, J., Daho, T., Vaitilingom, G., Koulidiati, J. (2020), "Comparative study of three ways of using Jatropha curcas vegetable oil in a direct injection diesel engine", Scientific African, v. 7, pp. e00290.

[11] Hazar, H., Sevinc, H., Sap, S. (2019), "Performance and emission properties of preheated and blended fennel vegetable oil in a coated diesel engine", Fuel, v. 254, pp. 115477.

[12] Dash, S. K., Lingfa, P. (2018), "Performance evaluation of Nahar oil–diesel blends in a single cylinder direct injection diesel engine", Int. J. Green Energy, v. 15, pp. 400-405.

[13] Raj, V. M., Subramanian, L. R. G., Manikandaraja, G. (2018), "Experimental study of effect of iso-butanol in performance, combustion and emission characteristics of CI engine fuelled with cotton seed oil blended diesel", Alexandria Eng. J., v. 57, pp. 1369-1378.

[14] Jain, N. L., Soni, S. L., Poonia, M. P., Sharma, D., Srivastava, A. K., Jain, H. (2017), "Performance and Emission Characteristics of Preheated and Blended Thumba Vegetable Oil in a Compression Ignition Engine", Applied Thermal Eng, v. 113, pp. 970-979.

[15] Karthik, N., Rajasekar, R., Siva, R., Mathiselvan, G. (2017), "Experimental Investigation of Injection Timing on the Performance and Exhaust Emissions of a Rubber Seed Oil Blend Fuel in Constant Speed Diesel Engine", Int. J. Ambient Energy, v. 40, pp. 282-294.

[16] Sathiyamoorthi, R., Sankaranarayana, G. (2017), "The effects of using ethanol as an additive on the combustion and emissions of a direct injection diesel engine fuelled with neat lemongrass oil-diesel fuel blend", Renew. Energy, v. 101, pp. 747-756.

[17] Sonar, D., Soni, S. L., Sharma, D., Srivastava, A., Goyal, R. (2015), "Performance and emission characteristics of a diesel engine with varying injection pressure and fuelled with raw mahua oil (preheated and blends) and mahua oil methyl ester", Clean Technol. Environ Policy, v. 17, pp. 1499-1511.

[18] Qi, D. H., Lee, C. F., Jia, C. C., Wang, P. P., Wu, S. T. (2014), "Experimental investigations of combustion and emission characteristics of rapeseed oil–diesel blends in a two cylinder agricultural diesel engine", Energy Convers Manag, v. 77, pp. 227-232.

[19] Agarwal, A. K., Dhar, A. (2013), "Experimental investigations of performance, emission and combustion characteristics of Karanja oil blends fuelled DICI engine", Renew. Energy, v. 52, pp. 283–291.

[20] Tamilselvan, P., Nallusamy, N., Rajkumar, S. (2017), "A comprehensive review on performance, combustion and emission characteristics of biodiesel fueled diesel engines", Renew Sustain Energy Rev, v. 79, pp. 1134-1159.

[21] Mahmudul, H. M., Hagos, F. Y., Mamat, R., Abdul Adam, A., Ishak, W. F. W., Alenezi, R. (2017), "Production, characterization and performance of biodiesel as an alternative fuel in diesel engines – A review", Renew Sustain Energy Rev, v. 72, pp. 497-509.

[22] Damanik, N., Ong, H. C., Tong, C. W., Mahlia, T. M., Silitonga, A. S. (2018), "A review on the engine performance and exhaust emission characteristics of diesel engines fueled with biodiesel blends", Environ Sci Pollut Res, v. 5, pp. 15307-25.

[23] Kathirvel, S., Layek, A., Muthuraman, S. (2016), "Exploration of waste cooking oil methyl esters (WCOME) as fuel in compression ignition engines: a critical review", Eng Sci Technol Int J, v. 19, pp. 1018-1026.

[24] Karakaya, H. (2019), "Effects of ethanol addition to biodiesel fuels derived from cottonseed oil and its cooking waste as fuel in a generator diesel engine", Energy Sources Part A Recovery Utilization and Environ Effects, v. 42, pp. 1359-1374.

[25] Hojati, A, Shirneshan, A. (2020), "Effect of compression ratio variation and waste cooking oil methyl ester on the combustion and emission characteristics of an engine", Energy Environment, v. 31, pp.1257-1280.

[26] Ors, I. (2020), "Experimental investigation of the cetane improver and bioethanol addition for the use of waste cooking oil biodiesel as an alternative fuel in diesel engines", J Brazil Soc Mech Sci Eng, v. 42, pp. 177.

[27] Al-Dawody, M., Jazie, A. A., Abbas, H. A. (2019), "Experimental and simulation study for the effect of waste cooking oil methyl ester blended with diesel fuel on the performance and emissions of diesel engine", Alexandria Eng J, v.58, pp. 9-17.

[28] Mathew, B. C., Thangaraja, J., Sivaramakrishna, A. (2019), "Combustion, performance and emission characteristics of blends of methyl esters and modified methyl esters of karanja and waste cooking oil on a turbocharged CRDI engine", Clean Technol Environ Policy, v. 21, pp. 1791–1807.

[29] Singh, D., Deep, A., Sadhu, S. S., Sharma, A. K. (2019)," Experimental Assessment of Combustion, Performance and Emission Characteristics of a CI Engine Fueled with Biodiesel and Hybrid Fuel Derived from Waste Cooking Oil", Environ Progress Sustain Energy, v. 38, pp. 13112.

[30] Wu, G., Jiang, G., Yang, Z., Huang, Z. (2019), "mission Characteristics for Waste Cooking Oil Biodiesel Blend in a Marine Diesel Propulsion Engine", Polish J. Environ Studies, v. 28, pp. 2911–2921.

[31] Yesilyurt, M. K. (2019), "The effects of the fuel injection pressure on the performance and emission characteristics of a diesel engine fueled with waste cooking oil biodiesel-diesel blends", Renew Energy, v. 132, pp. 649-666.

[32] Abed, K. A., El Morsi, A. K., Sayed, M. M., El Shaib, A. A., Gad, M. S. (2018), "Effect of waste cooking-oil biodiesel on performance and exhaust emissions of a diesel engine", Egypt J Petroleum, v. 27, pp. 985–989.

[33] Borugadda, V. B., Paul, A. K., Chaudhari, A. J., Kulkarni, V., Sahoo, N., Goud, V. V. (2018), "Influence of Waste Cooking Oil Methyl Ester Biodiesel Blends on the Performance and Emissions of a Diesel Engine", Waste Biomass Valor, v. 9, pp. 283:295.

[34] García-Martín, J. F., Barrios, C. C., Alés-Álvarez, F. J., Dominguez-Sáez, A., Alvarez-Mateos, P. (2018), "Biodiesel production from waste cooking oil in an oscillatory flow reactor. Performance as a fuel on a TDI diesel engine", Renew Energy, v. 125, pp. 546-56.

[35] Kumar, R., Gakkhar, R. P. (2018), "Influence of nozzle opening pressure on combustion, performance and emission analysis of waste cooking oil biodiesel fueled diesel engine", Int J Renew Energy Technol, v. 9, pp. 244–259.

[36] Ray, S. K., Prakash, O. (2018), "Biodiesel Extracted from Waste Vegetable Oil as an Alternative Fuel for Diesel Engine: Performance Evaluation of Kirlosker 5 kW Engine", In: Chattopadhyay, J., Singh, R., Prakash, O. (eds)., "Renewable Energies and Its Innovative Technologies, Springer, Singapore, pp. 219-229.

[36] Ulusoy, Y., Arslan, R., Tekin, U., Surman, A., Bolat, A., Sahin, R. (2018), "Investigation of performance and emission characteristics of waste cooking oil as biodiesel in a diesel engine", Petro Sci, v. 15, pp. 396–404.

[38] Wei, L., Cheng, R., Mao, H., Geng, P., Zhang, Y., You, K. (2018), "Combustion process and NOx emissions of a marine auxiliary diesel engine fueled with waste cooking oil biodiesel blends", Energy, v. 144, pp. 73-80.

[39] Geng, P., Mao, H., Zhang, Y., Wei, L., Yu, K., Ju, J., Chen, T. (2017), "Combustion characteristics and NOx emissions of a waste cooking oil biodiesel blend in a marine auxiliary diesel engine", Appl Thermal Eng, v.115, pp. 947-954.

[40] Meisami, F., Ajam, H., Tabasizadeh, M. (2017), "Thermo-economic analysis of diesel engine fueled with blended levels of waste cooking oil biodiesel in diesel fuel", Biofuel.

[41] Prabu, S. S., Asokan, M. A., Roy, R., Francis, S., Sreelekh, M. K. (2017), "Performance, Combustion and Emission Characteristics of Diesel Engine fueled with Waste Cooking Oil Bio-diesel or diesel blends with Additives", Energy, v. 122, pp. 638-648.

[42] Sabariswaran, K., Selvakumar, S. (2017), "Evaluation of engine performance and emission studies of waste cooking oil methyl ester and blends with diesel fuel in marine engine", Indian J Geo Marine Sci, v. 46, pp. 284-289.

[43] Wei, L., Cheung, C. S., Ning, Z. (2017), "Influence of waste cooking oil biodiesel on combustion, unregulated gaseous emissions and particulate emissions of a direct-injection diesel engine", Energy, v. 127, pp. 175-185.

[44] Yildizhan, S., Uludamar, E., Calik, A., Dede, G., Ozcanli, M. (2017), "Fuel properties, performance and emission characterization of waste cooking oil (WCO) in a variable compression ratio (VCR) diesel engine", Euro Mech Sci v. 1, pp. 56-62.

[45] Attia, A. M. A., Hassaneen, A. E. (2016), "Influence of diesel fuel blended with biodiesel produced from waste cooking oil on diesel engine performance", Fuel, v. 167, pp. 316-328.

[46] Kotebavi, V., Shetty, D., Sahu, D. (2016), "Performance and emission characteristics of a ci engine run on waste cooking oil-diesel blends", Pollution Res, v. 35, pp. 159-166.

[47] Chuah, L. F., Aziz, A. R. A., Yusup, S., Bokhari, A., Klemes, J. J., Abdullah, M. Z. (2015), "Performance and emission of diesel engine fuelled by waste cooking oil methyl ester derived from palm-olein using hydrodynamic cavitation", Clean Technol Environ Policy, v. 17, pp. 2229–2241.

[48] Capuano, D., Costa, M., Di Fraia, S., Massarotti, N., Vanoli, L. (2017), "Direct use of waste vegetable oil in internal combustion engines", Renew Sustain Energy Rev, v. 69, pp. 759-770.

[49] Pugazhvadivu, M., Jeyachandran, K. (2005), "Investigations on the performance and exhaust emissions of a diesel engine using preheated waste frying oil as fuel", Renew Energy, v. 30, pp. 2189–2202.

[50] Bari, S., Yu, C. W., Lim, T. H. (2002), "Filter clogging and power loss issues while running a diesel engine with waste cooking oil", Proceed Inst Mech Eng D J Auto Eng, v. 216, pp. 993–1001.

[51] Dhanasekaran, R., Krishnamoorthy, V., Rana, D., Saravanan, S., Nagendran, A., Rajesh Kumar, B. (2017), "A sustainable and eco-friendly fueling approach for direct-injection diesel engines using restaurant yellow grease and n-pentanol in blends with diesel fuel", Fuel, v. 193, pp. 419–431.

[52] Ravikumar, J., Saravanan, S. (2017), "Performance and emission analysis on blends of diesel, restaurant yellow grease and n-pentanol in direct-injection diesel engine", Environ Sci Pollut Res, v. 24, pp. 5381-5390.

[53] Krishnamoorthy, V., Dhanasekaran, R., Rana, D., Saravanan, S., Kumar, B.R. (2018), "A comparative assessment of ternary blends of three bio-alcohols with waste cooking oil and diesel for optimum emissions and performance in a CI engine using RSM", Energy Convers Manag, v. 156, pp. 337-357.

[54] Senthil Kumar, M., Jaykumar, M. A. (2014), "Comprehensive study on performance, emission and combustion behavior of a compression ignition engine fueled with WCO (waste cooking oil) emulsion as fuel", J Energy Institute, v. 87, pp. 263–271.

[55] D'Alessandro, B., Bidini, G., Zampilli, M., Laranci, P., Bartocci, P., Fantozzi, F. (2016), "Straight and waste vegetable oil in engines: Review and experimental measurement of emissions, fuel consumption and injector fouling on a turbocharged commercial engine", Fuel, v. 182, pp. 198–209.

[56] Hribernik, A., Kegl, B. (2009), "Performance and exhaust emissions of an indirect-injection (IDI) diesel engine when using waste cooking oil as fuel", Energy Fuels, v. 23, pp. 1754-1758.

[57] Kalam, M. A., Masjuki, H. H., Jayed, M. H., Liaquat, A. M. (2011), "Emission and performance characteristics of an indirect ignition diesel engine fueled with waste cooking oil", Energy, v. 36, pp. 397–402.

Chapter 3

TO DEVELOP A WASTE VEGETABLE OIL-BIODIESEL BASED DIESEL SUBSTITUTE FUEL BY RESPONSE SURFACE OPTIMIZATION OF DENSITY AND CALORIFIC VALUE

3.1 Introduction

The trans-esterified derivatives of vegetable oils have been widely explored as potential alternatives to diesel fuel. Utilizing WVO as a feedstock for producing biodiesel shall valorize WVO, a common liquid waste and also minimize environmental stress due to the disposal of WVO. Moreover, this strategy shall result in economic benefits in terms of waste management and also because of the production of cheaper biodiesel than vegetable oil biodiesel (due to the very negligible feedstock price of WVO). Several studies reported the evaluation of waste vegetable oil biodiesel (WVO-BD) for engine performance, combustion, and emission profiles [1-2].

The WVO-BDs are usually mixed in lower fractions to achieve properties of the blends close to those of diesel fuel. Lower fractions of WVO-BD in diesel blends are reported to show combustion characteristics similar to those of diesel in CI engines. However, reports exist where 0.20 to 0.50 WVO-BD fractions were blended with diesel [3-7]. In these cases, the densities and viscosities were higher for such blends. On the contrary, the calorific values of such blends were significantly reduced. These reports indicated a reduction in the engine performance (lower efficiency and higher fuel consumption compared to diesel) for the blends containing higher fractions of WVO-BD. Also, WVO-BD blends with diesel are reported to lower carbon monoxide (CO) and unburnt hydrocarbon (HC) emissions, but increased oxides of nitrogen (NO_x) emission [1, 2].

Reports on blending fuel oil having a lower density than diesel with straight vegetable oil biodiesels have been published in the literature [8-12]. Lighter fuel oil, like kerosene, acts as a viscosity reducer for biodiesel. Blending 0.50 fraction kerosene with straight vegetable oil biodiesel is reported to demonstrate physical properties similar to those of neat diesel fuel [8-10]. The engine performance of those kerosene blends with vegetable oil biodiesel was reported to be better (with lower fuel consumption and higher thermal efficiency) than diesel. The blends were reported to produce reduced CO emissions but increased HC and NO_x emissions relative to diesel. However, reports on blends of WVO-BD with kerosene are yet not available in the literature.

The use of fuel oil having a density in the gasoline range is restricted in CI engines due to higher volatility, increased ignition delay, and lower lubricity. Blending straight vegetable oil biodiesel with lighter fuel oil like gasoline is known to reduce the ignition delay, increase lubricity, and lower the volatility of the blend. The resultant blend was reported to be more suitable for utilization as a diesel alternative. Higher thermal efficiency and better combustion were reported for those blends in the literature [11, 12]. However, reports on the evaluation of biodiesel with lighter fuel oil like gasoline in a diesel engine are limited and not explored. The blends of WVO-BD with gasoline are yet to be evaluated in a diesel engine. Accordingly, the obvious research interest is to evaluate WVO-BD in blend with different fuel oil of varying density as an alternative fuel.

Literature suggests that external factors like temperature exposure can significantly affect the fatty acid composition of vegetable oil and its derivatives. The fatty acid composition of vegetable oil biodiesel governs the physical attributes that determine their performance, storage, and usability [13]. Thus, the temperature exposure of vegetable oil changes the level of unsaturation in fatty acids and the fatty acid composition. The changes in fatty acid composition lead to changes in the physical properties like density, viscosity, and calorific value of the resultant biodiesel [14]. Thus, the effects of varying thermal exposure duration of WVO on the properties of WVO-BD blends is a pertinent research scope.

Accordingly, the present study is planned to assess the properties of blend fuels of WVO-BD with different (fossil) fuel oils varying in terms of density. Another objective of the study is to understand the effect of the thermal exposure time of WVO on the property of blend fuel. The density of the fuel is an important physical property that defines the fuel economy in CI engines. Besides, the calorific value of fuel also impacts an engine's performance parameters and combustion characteristics. The mileage of fuel is primarily governed by the density and calorific value of the fuel [13]. Thus, another related scope is to evaluate the impact of the proportion of WVO-BD in blend with various fossil fuels on the properties of blended fuel. In addition, no study is available in the literature that deals with the prediction of the density of blend fuel with WVO-BD as an alternate fraction.

The present study aims to systematically evaluate the effects of different density fractions of fuel oils, the WVO thermal exposure time, and WVO-BD blend proportion on the density and subsequently on the calorific value of the WVO-BD blend fuel. There are several approaches for a systematic evaluation of the impact of multiple variables simultaneously. The response surface methodology (RSM) is one such widely used approach that can concurrently evaluate the effect of multiple independent factors on one or more dependent variables (response) and statistically optimize the factors through a well-designed study [15]. The present work plans to optimize the density and calorific value of blend fuel containing WVO-BD using the RSM technique. The blend fuel with optimized density and calorific value will be evaluated as a diesel alternative. A model for optimization of WVO-BD blend composition to design a diesel-fuel-alternative is a unique deliverable of the study. The study plans to outline a value-addition route of WVO-BD and define a real-life diesel fuel substitute.

3.2 Methodology

3.2.1 Materials

The fuel oil stocks were sourced from local fueling stations (Agartala, TR, India). The vegetable oil, of soybean origin, was used in the study. Analytical grade methanol of 99.5% purity (Loba Chemie Ltd., MH, India) was used in the study. The calcium oxide-based solid oxide catalyst was prepared using a method reported literature. In the method, as described by Majhi and Ray [16], waste eggshells were heated at 850±15 °C to obtain a calcium oxide-based solid oxide catalyst that had high basicity. The solid oxide catalyst, prepared from waste, is a low-cost substitute for commercial catalysts like sodium hydroxide or potassium hydroxide. Hence, the present process of WVO-BD synthesis should be more economic than commercial processes. The waste vegetable oil (WVO) is prepared by subjecting the vegetable oil to thermal treatment equivalent to food processing operation (of frying) for different time duration. Literature suggests that vegetable oils used for frying are subjected to thermal exposure at temperatures of 140-170 °C for a duration of not more than 72 hours [17]. Accordingly, the WVO stocks utilized in the study were obtained in-house by pretreating the vegetable oil stock at 150±5 °C for different durations that varied from 24 to 72 hours.

3.2.2 Preparation of waste vegetable oil-biodiesel (WVO-BD)

The free fatty acid content of the various WVO stocks was determined using the potassium hydroxide-based titration method. The free fatty acid content of the WVO stocks varied between 1.01 - 1.06 % (w/w). A free fatty acid content of less than 1.0 % (w/w) is recommended for utilizing the base-catalyzed transesterification process. Since the free fatty acids of the WVO stocks were close to 1.0 % (w/w), an additional pretreatment step, i.e., acid-catalyzed esterification was avoided to make the process cost-effective. This way the final cost of WVO-BD was maintained lesser than diesel. The transesterification of WVO was carried out using the methanol to oil molar ratio of 9:1 under continuous closed reflux for 2 hours at 65±5 °C with a 10 % (w/w) solid oxide catalyst. A higher methanol-to-oil ratio was reported to result in an increased yield of WVO-BD from the transesterification of WVO [16]. Since the objective of the present study was to use different transesterified WVO stocks as fuel blends, the condition that gives the maximum biodiesel yield was chosen. Accordingly, the methanol-to-oil ratio of 9:1 was used to produce different WVO-BD stocks. The principal reaction taking place during the transesterification process is shown below:

$$\text{Fatty acids or triglycerides} + 3\,CH_3OH \xrightarrow[65\pm5\ ^{\circ}C]{\text{Solid oxide Catalyst}} \text{Glycerol} + \text{Methyl esters (biodiesel)}$$

For the transesterification process, 100 ml (or 90 g) of WVO was mixed with 37 ml (or 29.3 g) of methanol which resulted in 103 ml (or 90.6 g) of WVO-BD along with 7.5 ml (or 9.4 g) of glycerol. Additionally, around 15 ml (or 11.9 g) of the excess methanol was distilled and collected at the end of the process. As a result, around 90% WVO-BD yield was recorded for the above-mentioned process. The WVO-BD was recovered from the reaction mixture after settling for 24h in a separating funnel wherein WVO-BD formed a distinct separable layer on top of the denser glycerol. The residual oil was trapped in the solid acid catalyst and remained at the bottom of the separating funnel. The recovered WVO-BD was washed repeatedly with water, heated at 105±5

°C to remove residual moisture content, and stored in hermetically sealed containers for characterization and subsequent blending.

3.2.3 Preparation of fuel blends and characterization

The blend fuel samples were prepared by direct volumetric blending of WVO-BD stocks with the fuel oils. Three different WVO-BD stocks (WVO-BD_24h, WVO-BD_48h, and WVO-BD_72h) were blended with three different fuel oils (of density 0.75 g/ml, 0.80 g/ml, and 0.83 g/ml) in three different blend ratios (0.20, 0.35, and 0.50) to prepare various blend fuel samples.

The various WVO-BD stocks and fuel oils of different density fractions were characterized by various fuel properties. The measurements of density, viscosity, cetane index, flash point, and fire point were performed as per ASTM standard test methods. The various test methods are discussed below, in brief:

The density of liquid fuel was determined as per ASTM D1217 test method. A 25 ml glass pycnometer (Make: Borosil Pvt. Ltd., India) which has a distinct 25 ml marking at the rim was selected. The fuel was carefully filled to the marked position to get a precise volume of 25 ml inside the pycnometer. After putting the lid on the container was kept inside an incubator at 15 °C where the weight of the container was measured using a high-precision digital balance. The empty mass of the pycnometer was subtracted from the filled mass to get the mass of 25 ml of the fuel. On dividing the mass of oil (in g) with the volume of oil (in ml), the density of the fuel (in g/ml) was determined.

For determining the viscosity of a fuel sample, a digital rotational viscometer (Model: DB-I Prime, Make: Brookfield Engg. Laboratories Inc., USA) was utilized and ASTM D2983 test method was followed. A water bath for retaining 40 °C temperature in the oil chamber and a specific spindle for low volume samples were selected. 16 ml oil sample was put inside the container along with the spindle which was attached to the meter with a freely suspended steel thread. Upon starting the motor, the spindle rotated inside the fuel the absolute viscosity (in centipoise) was displayed on the meter. Upon dividing the absolute viscosity with the density of the fuel at 40 °C (g/ml), the kinematic viscosity of the fuel (in mm²/sec) was determined.

The flash and fire points of fuel were determined using a Pensky-Martens closed cup apparatus (Make: Simeco Instruments Pvt. Ltd., India), conforming to the ASTM D93 method. Around 70 ml fuel sample was taken in the cup and the lid was attached. A digitally-controlled electric heater was used to heat the oil inside the closed cup and a stirrer ensured uniform heat distribution in the fuel. Fuel vapor was formed in the space above the oil surface. An external fire source was systematically lowered through a small opening on the lid such that it reaches the space above the oil surface. The temperature at which a sudden flash of fire didn't persist for more than 3 seconds was the flashpoint of the fuel. While the temperature at which a fire consistently burnt for more than 5 seconds on the oil's surface is considered as the fire point of the fuel.

The calorific value of the fuel was measured using a digital Bomb calorimeter (Model: BMB-100, Make: Petrotest Instruments Pvt. Ltd., India), following the ASTM D240 test method. In this regard, around 1.0 g of the fuel sample was weighed in the test cup and placed inside the Bomb, which was then filled with pure oxygen at the prescribed pressure (up to 30 kg/cm²). The bomb was placed in a water bath whose temperature was continuously monitored by a thermocouple. When the sample was combusted due to an electric ignition inside the Bomb, the rise in the temperature of the water was correlated to provide the gross calorific value of the fuel sample.

The cetane index of fuel was determined following ASTM D976 test method. Wherein, the temperature at which 50 % of the fuel was distilled out (known as T_{50}) was determined using an atmospheric distillation setup, conforming to the ASTM D86 method. The cetane index was determined using the following correlation:

$$Cetane\ index = 454.74 - 1641.416\ D - 774.74\ D^2 - 0.554\ T50 + 97.803\ (log\ T_{50})^2 \qquad (3.1)$$

Where D is the density of fuel at 15 °C and T_{50} is the temperature at which 50 % of the fuel was distilled out in an ASTM D86 atmospheric distillation process.

The fatty acid compositions of the various WVO-BD stocks were determined using a gas chromatography-mass spectrometry (GC-MS) (Clarus-600/560, Perkin Elmer, India) technique, as reported in the literature [7]. The oxidation stabilities of the various WVO-BD stocks were computed using a correlation reported in the literature [18]. The aliphatic, aromatic, and carbon content were also determined using the GC-MS technique [19].

3.2.4 Experimental design and modeling using RSM

3.2.4.1 Determination of design factors and levels

A feasibility study was performed to determine the associated levels for different experimental factors. The thermal exposure of vegetable oil in food processing operation rarely extend beyond 72 hours, average being around 48 hours [17]. Accordingly, the thermal exposure time of WVO was varied between 24 to 72 hours. The densities of the three fuel oils used in the study were: 0.830 g/ml, 0.800 g/ml, and 0.750 g/ml. These density fractions are representative of the nominal density of the commonly used fossil fuel types, namely, diesel, kerosene, and gasoline respectively. The fraction of WVO-BD in fossil fuel blend was varied among 0.20, 0.35, and 0.50. The WVO-BD fraction in blend beyond 0.50 was not considered as a conventional diesel engine that has not been reported to accommodate higher biodiesel fraction in blend fuel [1, 2]. Thus, an array comprising of 3 factors varied at 3 levels was selected as the experimental design for the optimization of the response. Table 3.1 lists the factors and their associated levels used for the experimental design.

Table 3.1 Factors and their associated levels in the experimental array.

Factor levels	WVO-BD fraction in the blend	Thermal exposure time of WVO (h)	Density of fuel oil (g/ml)
1	0.20	24	0.750
2	0.35	48	0.800
3	0.50	72	0.830

3.2.4.2 Development of the response surface model

Among the available 3 factor-3 levels experimental design, Box–Behnken design (BBD) is the most commonly selected experimental design procedure because of wider orthogonality, better predictability, and lesser experimental requirements. BBD is a three-level design based upon the combination of two-level factorial designs and incomplete block designs. BBD is a spherical design with excellent predictability within the spherical design space and requires fewer experiments than a full-factorial design with the same number of factors. Compared to the other design methods, the BBD technique is considered the most suitable for evaluating quadratic response surfaces particularly in cases when the prediction of response at the extreme level is not the goal of the model. In addition, the BBD technique is rotatable or nearly rotatable regardless of the number of factors under consideration, which is another advantage [20]. In BBD, blocks of various design settings are evaluated individually. Blocked designs are better designs if the design allows the estimation of individual and interaction factor effects independently of the block effects. This condition is called orthogonal blocking or simply "orthogonality". Blocks are assumed to have no impact on the nature and shape of the response surface. An orthogonally blocked or "orthogonal" response surface design has advantages over unblocked RS designs in terms of wider reach and predictability [15].

Accordingly, a BBD-based design was chosen to define the factor levels for the array of experiments used for response surface optimization (Table 3.1). A quadratic response surface (RS) model for the response (density of blend (g/ml)) was evaluated in the BBD-based experimental array. The RS model (Eq. (3.2)) expressed the density of blend as a function of the WVO fraction in the blend, thermal exposure time of WVO, and density of fuel oil (factors). In Eq. (3.2), B_0– B_9 are the regression coefficients for the model terms and ε is the "error" associated with the model.

Density of blend (g/ml) $= B_0 + B_1$(WVO-BD fraction in blend) $+ B_2$(thermal exposure time of WVO) $+ B_3$(density of fuel oil) $+ B_4$(WVO-BD fraction)$^2 + B_5$(thermal exposure time of WVO)$^2 + B_6$(density of fuel oil)$^2 + B_7$(WVO-BD fraction in blend)(thermal exposure time of WVO) $+ B_8$(WVO-BD fraction in blend) (density of fuel oil) $+ B_9$(density of fuel oil)(thermal exposure time of WVO) $+ \varepsilon$ (3.2)

Table 3.2 The response at different factor levels of the experimental array.

Exp. order	Experimental factors			Response (Density of blend		
	WVO-BD fraction in the blend	Thermal exposure of WVO (h)	Density of fuel oil (g/ml)	Rep 1 (g/ml)	Rep 2 (g/ml)	Rep 3 (g/ml)
1	0.20	24	0.800	0.806	0.805	0.806
2	0.50	24	0.800	0.833	0.834	0.833
3	0.20	72	0.800	0.810	0.812	0.808
4	0.50	72	0.800	0.841	0.842	0.839
5	0.20	48	0.750	0.789	0.782	0.785
6	0.50	48	0.750	0.816	0.821	0.818
7	0.20	48	0.830	0.830	0.834	0.835
8	0.50	48	0.830	0.855	0.853	0.853
9	0.35	24	0.750	0.802	0.800	0.798
10	0.35	72	0.750	0.811	0.809	0.807
11	0.35	24	0.830	0.839	0.838	0.842
12	0.35	72	0.830	0.846	0.846	0.848
13	0.35	48	0.800	0.818	0.816	0.820
14	0.35	48	0.800	0.818	0.819	0.822
15	0.35	48	0.800	0.821	0.826	0.822

The density (response) for each factor-level combination of the experimental array was determined and tabulated in Table 3.2. The measurements were performed in triplicates and each replicate was marked as a block. The replication helps to determine the experimental error in optimization.

The order and coefficients of the RS model were established through the analysis of variance (ANOVA) of the response tabulated in Table 3.3. The RS model coefficients for the response were established from multiple regression analysis and determination of statistical significance at a 95% confidence level. A backward elimination method was employed by deleting the statistically insignificant terms to define the RS model. All statistical computations were performed using MINITAB® v. 16 statistical software (Minitab Inc., State College, PA, USA). Additional experiments were conducted and measurements were made in triplicates to validate the RS model and confirm the model accuracy.

3.2.5 Prediction of the calorific value of WVO-BD blends and validation of the prediction

Literature established the calorific value of fuel as a function of fuel density. Separate relations exist correlating the calorific value with the density of fuel oil and biodiesel [21, 22]. The reported relations were combined herein and used to predict the calorific value of the fuel blends from the RS model-predicted blend density. Additional experiments were performed to validate the prediction of the calorific value of the blend from the RS model-predicted density of the blended fuel.

3.2.4 Evaluation of optimized WVO-BD blend in a CI engine

The optimized blend fuel defined by the RS model-based density and the calorific value was experimentally prepared by blending the applicable WVO-BD stock in optimized proportion with the appropriate fuel oil stock. The experimentally reproduced optimized blend fuel (hereafter referred to as the model-optimized blend) was evaluated for the fuel properties. The characterized blend was assessed in a conventional compression ignition (CI) engine as illustrated in Fig. 3.1.

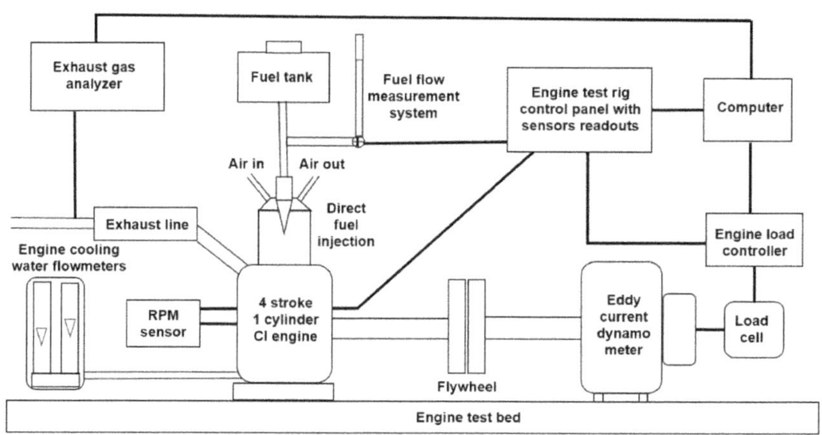

Fig. 3.1 Experimental test setup.

Table 3.3 Specifications of the engine setup and exhaust gas analyzer.

Parameters	Specifications
Engine	Kirloskar Oil Engines Ltd., Model – AV1
Type	Vertical, stationary, four-stroke, single-cylinder, naturally aspirated, direct injection, compression ignition
Bore x stroke	80 mm x 110 mm
Compression ratio	16.5: 1
Swept volume	0.553 liters
Rated speed	1500 RPM
Rated power	3.7 kW
Injection pressure	210 bar
Injection timing	23° before Top Dead Center
Starting mechanism	Hand cranking
Cooling system	Cooling water
Dynamometer	Powermag Control Systems Ltd., Model - FTAC
Mechanism	Eddy current
Exhaust gas analyzer	AVL India Ltd., Model – DiGas 444

A 3.7 kW, 4-stroke, single-cylinder, naturally aspirated, water-cooled, direct injection CI engine (AV-1, Kirloskar Oil Engines Ltd., MH, India) was operated with different fuel at the rated speed of 1500 RPM for the performance analysis. Through the operation, the compression ratio, fuel injection pressure, and injection timing were maintained invariant at 16.5:1, 210 bar, and 23° before the top dead center, respectively. All due care, such as flushing of the fuel line, proper engine warm-up, and equilibration of the engine, was strictly enforced before the test run with the experimental fuel. Through the operation, cooling water was continuously supplied to the engine to avoid over-heating. The engine performance parameters were recorded for low and high engine load conditions, corresponding to 20% and 80% of full engine load, respectively. The engine loads were fixed using an eddy-current type dynamometer (FTAC, Powermag Control Systems Ltd., India). The specifications of the engine setup used in the study are detailed in Table 3.3. The brake-specific fuel consumption (BSFC), brake thermal efficiency (BTE), mechanical efficiency (ME), and volumetric efficiency (VE) were determined for the model optimized blend and diesel in the engine performance analysis.

The optimized blend was evaluated for commonly studied emission parameters to characterize the emission profile. Accordingly, carbon monoxide (CO), carbon dioxide (CO_2), unburnt hydrocarbon (HC), and oxides of nitrogen (NO_x) emission of the blend fuel and diesel were determined from exhaust gas analysis during engine operation. The engine emissions were measured using an exhaust gas analyzer (DiGas 444, AVL India Ltd., MH, India). The resolution, accuracy, and uncertainties in emission measurements are tabulated in Table 3.4. The overall uncertainty in emission measurement was obtained to be ± 0.378%.

Table 3.4 Uncertainties associated with the measurement of engine emissions.

Measurement	Measurement technique	Working range	Resolution	Accuracy	% Uncertainty
Carbon monoxide	NDIR principle	0-0.10 % vol.	0.01 % vol.	± 0.03 % vol.	± 0.20
Carbon dioxide	NDIR principle	0-0.20 % vol.	0.01 % vol.	± 0.5 % vol.	± 0.15
Hydrocarbon	NDIR principle	0-2000 ppm	1 ppm	± 10 ppm	± 0.20
Nitrogen oxide	Electrochemical	0-5000 ppm	1 ppm	± 20 ppm	± 0.20

Note: NDIR – Non-dispersive Infrared

3.3 Experimental findings

3.3.1 Characterization of fuel stocks

The various WVO-BD stocks were characterized by fatty acid composition. The complete fatty acid methyl ester profile of the three WVO-BD stocks is presented in Table 3.5. All WVO-BD stocks have a high compositional presence of methylated esters of polyunsaturated fatty acids (PUFA) (linoleic acid (C18:2) and linolenic acid (C18:3)), followed by methylated esters of monounsaturated fatty acids (MUFA) (oleic acid (C18:1)). Palmitic acid (C16:0) and stearic acid were the abundant methylated esters of saturated fatty acids (SFA) in WVO-BD stocks. The fatty acid methyl ester profiles of different WVO-BDs are in agreement with the literature [23]. The total methylated esters of unsaturated fatty acid were noted to marginally decrease with an increase in the thermal exposure duration. The decrease in the number of methylated esters of unsaturated fatty acids content can be expected to affect the oxidation stability of biodiesel [18].

Table 3.5 Fatty acid methyl ester composition of WVO-BD prepared from varying thermally exposed WVO.

Fatty acid	MW (g/mol)	Carbon structure	Formula	WVO-BD-24h (wt%)	WVO-BD-48h (wt%)	WVO-BD-72h (wt%)
Methyl myristate	242	C14:0	$C_{15}H_{30}O_2$	0.05	0.02	0.02
Methyl palmitate	270	C16:0	$C_{17}H_{34}O_2$	13.17	13.82	14.01
Mehtyl stearate	298	C18:0	$C_{19}H_{38}O_2$	9.84	10.31	10.58
Methyl oleate	296	C18:1	$C_{19}H_{36}O_2$	34.67	34.23	35.09
Methyl linoleate	294	C18:2	$C_{19}H_{34}O_2$	41.23	40.07	38.84
Methyl linolenate	292	C18:3	$C_{19}H_{32}O_2$	1.46	1.12	1.08
Total SFA				23.06	24.15	24.61
Total MUFA				33.67	34.23	35.07
Total PUFA				42.69	41.19	39.82
Free fatty acid				0.09	0.07	0.06

Note: MW – molecular weight
SFA - saturated fatty acid (CX:0)
MUFA – monounsaturated fatty acid (CX:1)
PUFA – polyunsaturated fatty acid C(X:2), and C(X:3).
where, X denotes carbon number; 0, 1, 2, and 3 denotes the respective number of double bonds in the carbon chain.

The different fuel stocks were characterized for fuel properties. The results of the characterization are presented in Table 3.6. The density, viscosity, and flash point of WVO-BD stocks were observed to increase with an increasing thermal exposure time of WVO. The density and viscosity of WVO-BD varied between 0.873-0.887 g/ml and 4.61-4.74 mm^2/s, respectively. Increased thermal exposure and resultant reduced volatility were observed to increase the flash point and fire points for WVO-BD stocks. The increment in density, viscosity, flash point, and fire point for WVO-BDs with extended thermal exposure may be attributed to the loss of unsaturation and formation of polymerization products in WVO feedstock used for transesterification to WVO-BD

[17]. However, the properties of all the WVO-BD stocks used in the study conformed to the ASTM D6751 standard for biodiesel.

The oxidation stabilities of the WVO-BD stocks were noted to increase slightly with thermal exposure duration. The oxidation stabilities of the three WVO-BD were 4.45h, 4.67h, and 4.71h. The oxidation stabilities of the WVO-BD stocks complied with the ASTM D6751 standard for biodiesel (3h) but were non-compliant with the EN 14112 standard of 6h. A decrease in the PUFA content in WVO-BD with increased thermal exposure of feedstock WVO can be attributed to the marginal change in oxidation stability [18]. The observed oxidation stability values were in agreement with the values reported in the literature [24]. But, the oxidation stabilities of WVO-BD were noted to be significantly lower than the literature-reported values of straight vegetable oils. Largely, increased thermal exposure duration was found not to alter the storage stability and usability of the WVO-BD stocks.

Table 3.6 Properties of different fuel stocks used in the study.

Fuel samples	Fuel oil-1	Fuel oil-2	Fuel oil-3	WVO-BD-24h	WVO-BD-48h	WVO-BD-72h	Test method
Carbon content (% wt.)	86.6	86.5	86.5	76.6	76.8	76.9	GC-MS [19]
Aliphatic content (% wt.)	30.6	35.4	50.1	99.96	99.95	99.95	GC-MS [19]
Aromatic content (% vol.)	25.8	15.6	10.2	nd.	nd.	nd.	GC-MS [19]
Density at 15°C (g/ml)	0.823	0.798	0.754	0.873	0.881	0.887	ASTM D1217
Viscosity at 40°C (mm^2/s)	3.38	1.91	0.62	4.61	4.67	4.74	ASTM D2983
Flash point (°C)	48	46	---	131	136	139	ASTM D93
Fire point (°C)	58	51	---	>200	>200	>200	ASTM D93
Calorific Value (MJ/kg)	46.65	46.98	47.45	41.06	40.97	40.81	ASTM D240
Cetane index	51.25	48.38	---	55.32	55.14	55.46	ASTM D976
Oxidation stability[*] (h)	---	---	---	4.51	4.67	4.71	[18]

Note: nd. – not detected. [*]– Oxidation stability = 22.318 – 0.234*(linoleic acid wt%) [18].

A comparative presentation of the properties of the different fuel oils used in the study showed that viscosities and flashpoints of all fuel oils were lower, while the densities and calorific values were higher than the WVO-BD stocks used in the study. The physical properties of the fuels are dependent on their chemical composition. The fuel oils, having fossil origin, recorded higher carbon content (~86%) than WVO-BD stocks (~77%). The higher cetane indices of the WVO-BDs than the fuel oils indicate that WVO-BD blends can reduce the ignition delay in a CI engine, compared to neat diesel fuel. The dissimilar density, volatility, and energy content of the fuel oils compared to the WVO-BD stocks establish the very purpose of blending proposed in the study. WVO-BD with a higher aliphatic content can prove very effective upon blending with considerably aromatic fuel oil. Thus, the blending of various fuel oils with different WVO-BD stocks in varied proportions is expected to produce a fuel blend that can substitute diesel fuel. The WVO-BD stocks were noted to be completely miscible in the fuel oils in all proportions. Hence, the blending of all WVO-BDs with the fuel oils in different blend fractions resulted in the formation of consistent blends, without phase separation.

3.3.2 Development of response surface (RS) model

The impacts of various experimental factors on the density of the blend fuels were evaluated individually and collectively. The individual and collective factor effects are presented in Fig. 3.2. The analysis of individual factors (Fig. 3.2(a)) revealed the dominant influence of the fuel oil density and WVO-BD proportion on the blend density. The collective factor study (Fig. 3.2(b)) showed evidence of strong factor-level interaction for all three experimental factors. The interactions at various factor-level combinations for different experimental factors justified the choice of a non-linear quadratic factor-response relation considered in the study.

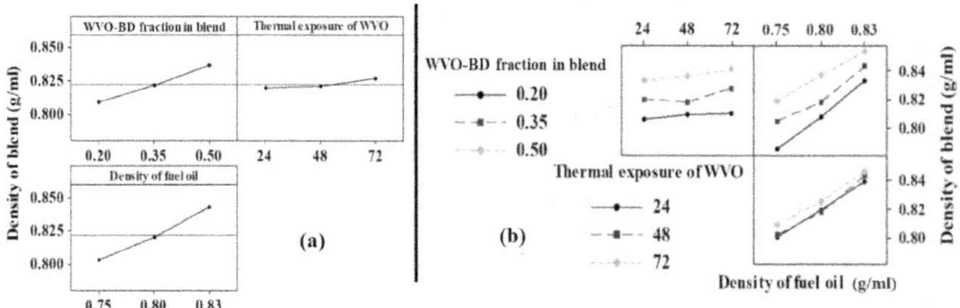

Fig. 3.2 (a) Individual factor effect plots for various experimental factors; **(b)** two-factor-at-a-time interaction plots for different experimental factors.

The quadratic response surface (RS) model (proposed in Eq. (3.1)) was evaluated using analysis of variance (ANOVA) for the measured density (response) for each factor-level combination of the experimental design tabulated in Table 3.2. ANOVA was used to determine the order of the RS model and define the regression coefficients (B_0– B_9) of the model terms. The ANOVA results, presented in Table 3.7, revealed that the model is statistically significant with linear, quadratic, and interaction terms. The significance of the model terms was determined from the p-values associated with the model terms.

A lower p-value (p = 0.0000) for overall regression established that the RS model is statistically significant at a 95% confidence level (p < 0.05) and confirmed a strong correlation between the model and the experimental response. A p-value (p = 0.685) associated with the lack-of-fit term confirmed that the error associated with the model is statistically insignificant (p > 0.05) at a 5% level of significance and the model has a good fit with the experimental response.

Table 3.7 Results of the analysis of variance of the RS model.

Source		DF	Adj. MS	F	P
Blocks		2	0.00000	0.3	0.742
Regression		9	0.00166	351.6	0.000*
Linear		3	0.00025	51.8	0.000*
	WVO-BD fraction in blend	1	0.00011	23.6	0.000*
	Thermal exposure time of WVO	1	0.00000	0.0	0.837
	Density of fuel oil	1	0.00058	123.2	0.000*
Square		3	0.00025	53.3	0.000*
	(WVO-BD fraction in blend)^2	1	0.00003	5.9	0.021*
	(Thermal exposure time of WVO)^2	1	0.00009	18.9	0.000*
	(Density of fuel oil)^2	1	0.00069	146.4	0.000*
Interaction		3	0.00004	7.9	0.000*
	WVO-BD fraction in blend x Thermal exposure time of WVO	1	0.00001	1.4	0.240
	WVO-BD fraction x Density of fuel oil	1	0.00010	21.3	0.000*
	Thermal exposure time of WVO x Density of fuel oil	1	0.00001	1.0	0.326
Residual		33	0.00001		
	Lack-of-Fit	27	0.00001	0.80	0.685
	Pure Error	6	0.00000		
Total		44			

Notes: DF = degrees of freedom; MS = adjusted mean of square; (*) = values are statistically significant at 5% level of significance; *p* = a statement describing F.

Table 3.8 Regression coefficients for the terms of the RS model.

Terms	Reg. coeff.	SE	T	P
Constant	3.57678	0.2768	12.9240	0.0000*
WVO-BD fraction in blend	0.41150	0.0848	4.8550	0.0000*
Thermal exposure time of WVO	0.00262	0.0126	0.2070	0.8370
Density of fuel oil	(-) 7.73963	0.6972	(-) 11.1010	0.0000*
(WVO-BD fraction in blend)^2	0.07037	0.0290	2.4260	0.0210*
(Thermal exposure time of WVO)^2	0.00283	0.0007	4.3410	0.0000*
(Density of fuel oil)^2	5.34028	0.4413	12.1000	0.0000*
WVO fraction in blend x Thermal exposure time of WVO	0.00500	0.0042	1.1960	0.2400
WVO-BD fraction in blend x Density of fuel oil	(-) 0.47475	0.1029	(-) 4.6130	0.0000*
Thermal exposure time of WVO x Density of fuel oil	(-) 0.01540	0.0154	(-) 0.9980	0.3260

Note: p = a statement which describes T; Reg. coeff. = regression coefficient; SE = Sequential sum of squares; (*) = values are statistically significant at 5% level of significance.

The computed regression coefficients for the model terms along with their respective p-values for the full quadratic RS model are presented in Table 3.8. A backward elimination method was applied and statistically insignificant terms (p > 0.05) were deleted from the full quadratic model to obtain a final RS model. The refined RS model significant at a 95% confidence level is presented as Eq. (3.3).

Density of = (3.57678) + (0.41150*WVO-BD fraction in the blend) - (7.73963*Density of fuel   (3.3)
blend (g/ml)   oil) + (0.07037*WVO-BD fraction in the blend*WVO-BD fraction in the blend) + (0.00283*Thermal exposure time of WVO*Thermal exposure time of WVO) + (5.34028*Density of fuel oil*Density of fuel oil) - (0.47475*WVO-BD fraction in the blend*Density of fuel oil)

The percent contribution and cumulative contribution of different model terms on the total model variance are presented as a Pareto chart in Fig. 3.3. The density of fuel was found to have contributed close to 36%, while its quadratic form showed more than 42% impact on the density of blend. The two terms, combined, contributed close to 78% to the response of the RS model. Also, the WVO-BD fraction had more than a 7% impact on the response. The quadratic term for the thermal exposure time of WVO was found to have impacted the response by 5.7%. Also, the interactions were particularly noticeable among WVO-BD blend fractions and the density of the fuel oil, with a contribution of 6.4% to the response. The most model variance (> 97%) can be explained by the above five model terms (Fig. 3.3).

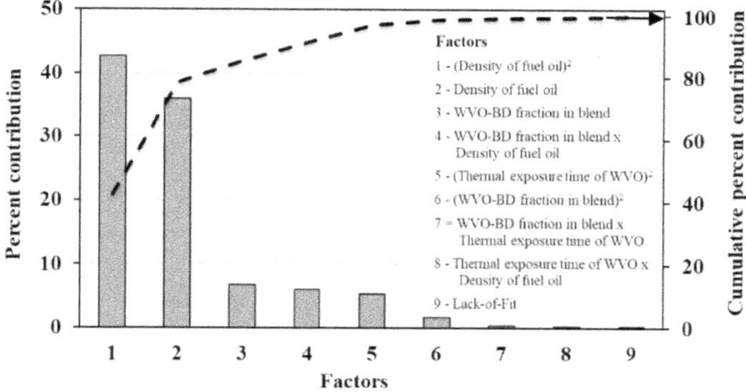

Fig. 3.3 Percent contribution and cumulative contribution of different model terms on the variance of the model for predicting the density of blend (Pareto chart).

3.3.3 Validation of the RS model

The different WVO-BD blend densities, determined from the quadratic RS model, were plotted against the densities of the equivalent WVO-BD blends prepared experimentally. The plot of the densities of experimental blends against the density of the equivalent predicted by the model is shown in Fig. 3.4(a). The R^2 value of 0.904 (corresponding to an R-value of 0.951) suggested that the model prediction was well correlated with the experimental values. The verification of the model prediction was achieved by analysis of the residuals. The residuals between the model-predicted and the experimental values were further subjected to the Anderson-Darling (AD) normality test to confirm that the residuals were non-systematic (random) in nature [15]. The percent probability plot of the residual between the model-predicted and the experimental values with related AD statistics is reported in Fig. 3.4(b). AD statistic value of 0.267 (shown in the plot) being much lower than the tabulated critical value of 0.752 and associated p-value of 0.672 (shown in the plot) being much greater than 0.05 confirmed that the difference between model prediction and experimental values are purely random in nature. A paired t-test for the entire dataset of model prediction and experimental values was performed to determine whether the values are equivalent [15].

The paired t-test confirmed that the difference was not statistically significant since the t-calculated (0.04) < t-tabulated (2.92)) at a 95% confidence level.

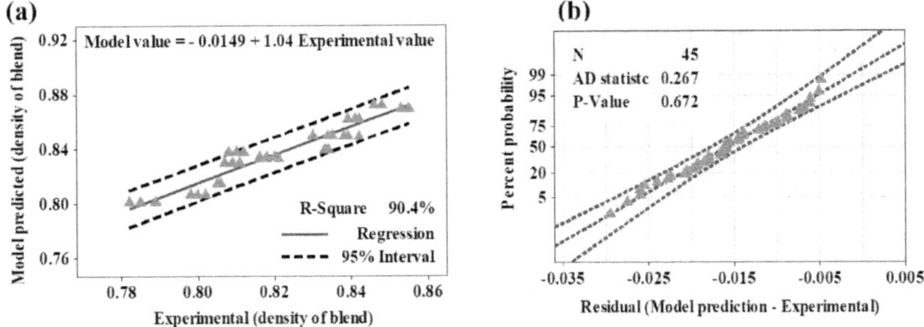

Fig. 3.4 (a) Correlation plot of experimental versus Model-predicted density of blend, **(b)** Anderson-Darling normality plot for prediction of the density of blend.

For the validation of the RS model, additional blends were prepared and the densities of the blends were determined. The experimental values of the density of blend were compared with model-predicted values for similar factor-level settings. The comparative plot of the densities of the experimental blend against model-predicted density for different WVO-BD fractions, thermal exposure time of WVO, and density of fuel oil are presented in Fig. 3.5. The closeness of model prediction against experimental values for each factor level was confirmed from the comparison of the t-statistic [15]. The computed t-values at all factor levels were found lower than the tabulated t-value for a 95% confidence level. Accordingly, the model-predicted values were deemed equivalent to the experimental values and the model for the density of blends was confirmed to be valid for different factors at all the factor levels considered in this study.

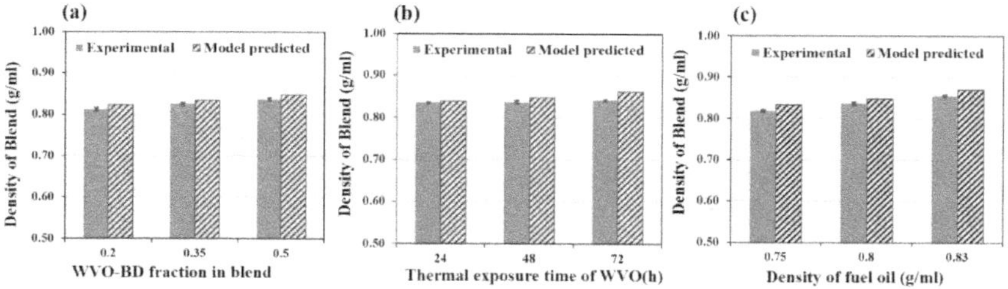

Fig. 3.5 Validation study of the RS model for **(a)** different WVO-BD fractions in the blend **(b)** various WVO thermal exposure times, and **(c)** the density of fuel oil.

3.3.4 Study of factor effect and response optimization

The validated RS model was used to analyze the impact of the experimental factors on the model response (density of blend (g/ml)) over the entire factor space. Fig. 3.6 represents the various surface plots for different factors (considering two-factor-at-a-time). The surface plot for WVO-BD fraction in the blend and the density of fuel oil (Fig. 3.6(a)) showed the lowest density of fuel oil (0.75 mg/l) coupled with the lowest WVO-BD fraction (0.20) resulted in the lowest density of blend. The density of blend increased with the increase in WVO-BD fraction and density of the fuel oil. Accordingly, mid to high WVO-BD fraction blended with mid to high fuel oil density considered can be expected to produce a blend with a density equivalent to diesel. The increase in the thermal exposure time of WVO was noted to have little effect on the increase in the density of the blend (Fig. 3.6(b)). The blend of mid to high-density fuel oil with WVO-BD stocks resulting from WVO with low to medium thermal exposure time was predicted as the favorable composition to produce diesel equivalent blend fuel. Similarly, Fig. 3.6(c) revealed that with a boost in the WVO-BD fraction, the blend density increased accordingly. Whereas, the thermal exposure time of WVO had little role in the change. Thus, 0.35 to 0.50 WVO-BD fraction, 0.80 to 0.83 g/ml fuel oil density, and 24 to 48 hour thermal exposure time of WVO was identified as the ideal range of experimental factors for the desired response, that is, blend having density equivalent to the diesel.

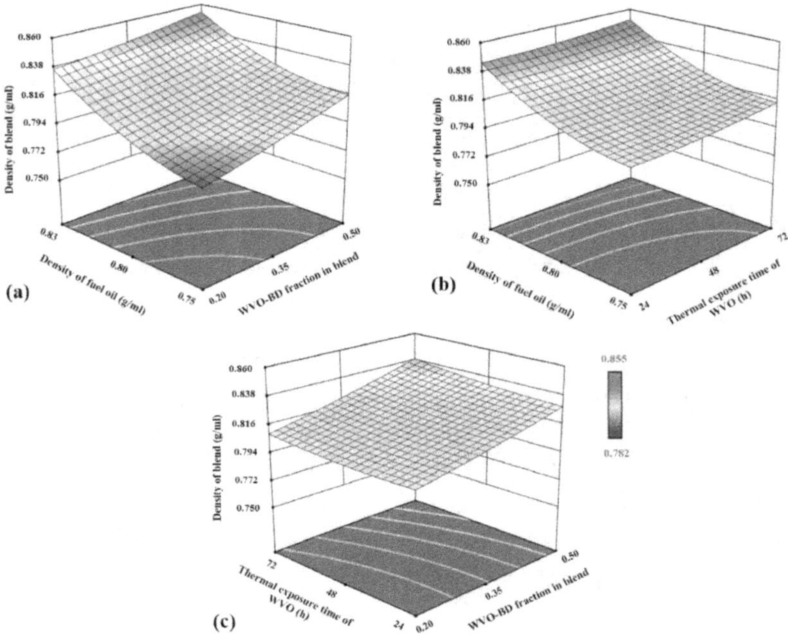

Fig. 3.6 Surface plots of the blend density for different experimental factors (two-at-a-time) **(a)** WVO-BD fraction in the blend and density of fuel oil, **(b)** Thermal exposure time of WVO and density of fuel oil, and **(c)** Thermal exposure time of WVO and WVO-BD fraction in the blend.

Further optimization of the experimental factor settings for the desired response was performed using the overlaid contour plots (Fig. 3.7). The contour plot in Fig. 3.7(a), showed that desirable factor-level setting of WVO-BD fraction and density of fuel oil to generate diesel equivalent blend density. As the density range for diesel fuel in the standard (IS 1460) varies between 0.810 to 0.845 g/ml [35], so the desired density of blend was fixed between 0.820 g/ml and 0.830 g/ml. A lower WVO-BD fraction was noted to accommodate a narrower band of fuel oil density to obtain the desired bend density. Higher density fuel oil was noted to blend with a lower WVO-BD fraction to produce desired blend density. For 0.35 WVO-BD fractions in the blend, the fuel oil density should be between 0.81 – 0.82 g/ml. WVO-BD can be maximum utilized at 0.50 fraction in blend with fuel oil having a density between 0.76 to 0.80 g/ml to produce desired blend density (0.82–0.83 g/ml). The fuel oil having a density between 0.76 to 0.80 g/ml can be blended with WVO-BD prepared from WVO with 48 hours of thermal exposure time (Fig. 3.7(b)) to produced desired blend fuel density. The range of fuel oil density narrowed between 0.77 to 0.80 g/ml as thermal exposure of WVO decreased below 48 hours. For increased thermal exposure of WVO beyond 48 hours, the desired fuel oil density region was shifted from 0.75 to 0.77 g/ml. A blend fraction of 0.37 to 0.42 across all thermal exposure times of WVO was noted to produce the desired blend fuel density (Fig. 3.7(c)). WVO-BD stock from 48 hours thermal exposure of WVO can be blended at a blend fraction of 0.37 to 0.48. The same blend fraction was applicable for thermal exposure time lower than 48 hours. But, higher thermal exposure of WVO, above 48 hours lowered the blending fraction between 0.32 and 0.41 to attain the desired density range of blend fuel.

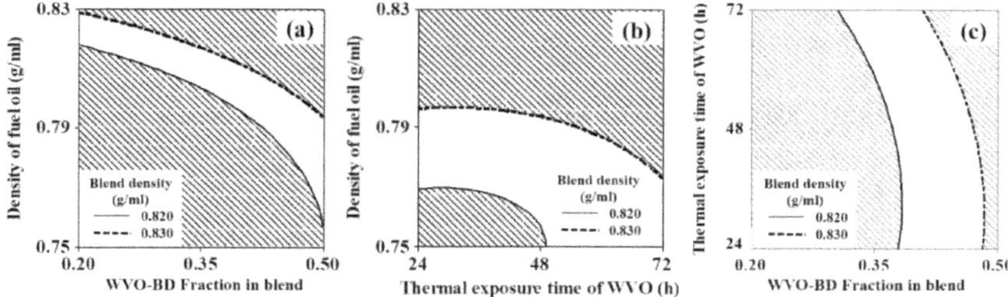

Fig. 3.7 Overlaid contour plots showing the desirability region- **(a)** for various WVO-BD fractions in the blend and density of fuel (g/ml), **(b)** for the density of fuel (g/ml) and thermal exposure time of WVO (h), and **(c)** for the thermal exposure time of WVO (h) and the fraction of WVO-BD in the blend.

The D-optimality criterion was employed to determine the model-optimized blend density based on the three factors considered in the study. A D-optimality of 0.9531 was achieved as the highest optimization outcome with a blend of density 0.827 g/ml, as shown in Fig. 3.8. The optimum blend density (0.827 g/ml) was predicted for a blend of WVO-BD prepared from WVO with 48 hours of thermal exposure with fuel oil of density 0.80 g/ml at a blend fraction of 0.46. An equivalent blend when prepared experimentally was noted to have a density of 0.829±0.004 g/ml. The error of prediction was computed to be less than 1.0% between the density model-optimized blend and equivalent experimental blend.

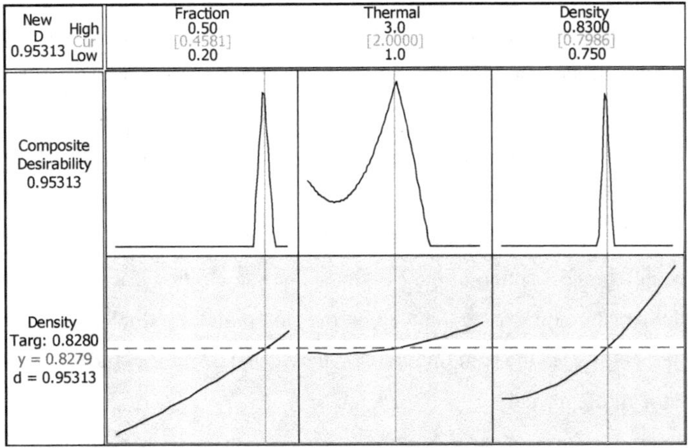

Fig. 3.8 Optimization plot for prediction of the density of the blend.

3.3.5 Determination of calorific value of blend fuel based on the model-predicted density

The calorific values of the blended fuel can be determined from the density of the WVO-BD blend. Accordingly, the calorific value of blend fuel was predicted from blend density using the relation shown in Eq. (3.4). The relation (Eq. (3.4)) was derived from the relation reported in the literature [33, 34].

Calorific value of blend (MJ/kg) = {(WVO-BD fraction in blend)*(63.776 − 25.90*Density of BD(g/ml))} + {(1 − (WVO-BD fraction in blend))*(55.75 − 11.051*Density of fuel oil(g/ml))}   (3.4)

The model predicted blend densities were used to compute the calorific values of the blend fuels using Eq. (3.4). The accuracy of the model-based prediction of calorific value was validated for different experimental factors over the entire factor-level settings defined in the study. Additional blends were prepared and the densities of the blends were predicted using the RS model. The calorific values of the blends were subsequently determined experimentally. The experimental calorific values of the blend were compared with the calorific values predicted from blend density for similar factor-level settings. The comparative plot of the experimental calorific values against predicted calorific values for different experimental factors is presented in Fig. 3.9. The experimental calorific values of blends were noted to be in close agreement with the calorific values predicted from the density of the blend defined by the RS model over various WVO-BD blend fractions (Fig. 3.9(a)). A similar agreement between experimental and predicted calorific values was also noted for the different thermal exposure durations of WVO (Fig. 3.9(b)). Fig. 3.9(c) revealed that the prediction of calorific values was also valid for the entire range of fuel oil densities evaluated in the study. The differences between experimental calorific values and predicted calorific values were statistically compared (using t-test) to confirm the validation [15]. The statistical comparison established that the experimental values and model-predicted values were statistically similar to each other (t-computed < t-tabulated) for different experimental factors across all factor-level settings evaluated. Hence, the prediction relation for calorific values, given by Eq. (3.3), can be considered valid over the entire factor-space defined in the study.

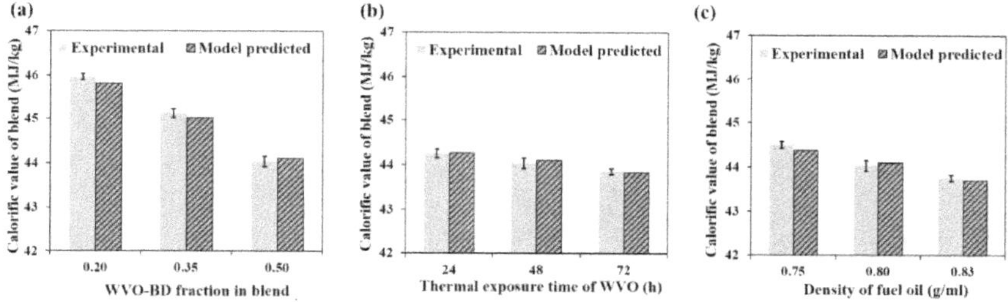

Fig. 3.9 Validation of the predicted calorific values against experimental calorific values of the blend fuel for the factors under consideration - **(a)** WVO-BD fraction in the blend, **(b)** Thermal exposure time of WVO (h), and **(c)** and Density of fuel oil (g/ml).

The validated prediction relation for calorific values of the blend fuel was used to predict the optimum experimental factor settings for the desired region of the calorific value of blend fuel. The overlaid contour plots for the desired region of the calorific value of the blended fuel for different experimental factors are shown in Fig. 3.10. The calorific value of the standard diesel fuel was identified as the desired region for the calorific value of the blended fuel. The calorific value of standard diesel fuel reportedly ranged between 43–47 MJ/kg. Accordingly, the desirable region of the calorific value for blend fuel was kept between 44 MJ/kg and 46 MJ/kg for the identification of optimum settings for different experimental factors. The higher thermal exposure time of WVO beyond 48 hours was noted to have a lower WVO-BD fraction in the blend for the desirable calorific value of the resultant blend fuel (Fig. 3.10(a)). The relation predicted that the thermal exposure time of WVO of 48 hours and below can be blended at 0.50 blend fraction with the desirable calorific value of the blended fuel. Higher fractions of WVO-BD can be blended with fuel oil of lower density to achieve a desirable calorific value (Fig. 3.10(b)). The relation predicted that fuel oil of density 0.80 g/ml or below can be blended up to 0.50 blend fraction within the desirable range of calorific value. Further noted that the fuel oil density of 0.80 g/ml can accommodate all the WVO-BD fractions from 0.20 to 0.50 within the desirable range of calorific value of the blend fuel.

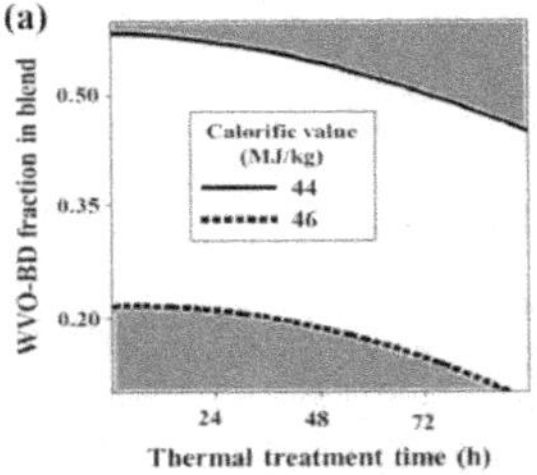

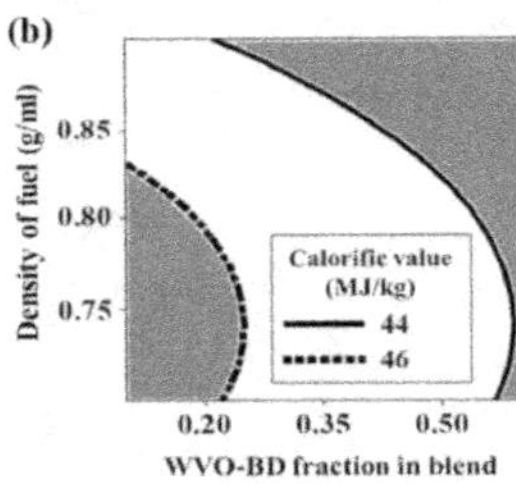

Fig. 3.10 Overlaid contour plots showing the desirability region for calorific value of the blend - **(a)** various thermal exposure time of WVO (h) versus WVO-BD fractions in the blend, and **(b)** WVO-BD fractions in the blend versus the density of fuel oil (g/ml).

Accordingly, a calorific value of 44.5 MJ/kg was predicted for an optimum blend of WVO-BD prepared from WVO with 48 hours of thermal exposure with fuel oil of density 0.80 g/ml at a blend fraction of 0.46. An equivalent blend when prepared experimentally was noted to have a calorific value of 44.0±0.5 MJ/kg. The error of prediction was computed to be less than 1.0% between the predicted calorific value of blend and the experimental calorific value of equivalent blend.

3.3.6 Comparison of the model-optimized blend fuel against diesel

The blend fuel with optimized composition predicted by the model was experimentally prepared and evaluated for fuel properties. The experimentally prepared blend fuel with model optimized composition was designated hereafter as the model-optimized blend. The properties of the model-optimized blend were compared against diesel to establish the relevance of the blend as a diesel alternative. The comparison of fuel properties of the model-optimized blend against diesel is presented in Table 3.9. The density of the optimized blend was comparable to diesel. A t-test revealed that the difference between the densities of the blended fuel with that of diesel was statistically insignificant at a 5% level of significance (t-computed (2.08) < t-tabulated (3.18)). The results confirmed that the RS-based modeling approach had been successful in defining a diesel alternative with equivalent density. However, the model-optimized blend was noted to have viscosity than diesel. Better flow properties and fuel atomization can be expected for the blend in comparison to diesel. Also, the flash and fire point of the blend was observed to be higher than that of diesel. The higher flash and fire point make the blend safer than diesel for handling and storage. The cetane index of the blend was found comparable to that of diesel. Accordingly, the least variation of ignition delay can be expected on substituting diesel with the blended fuel in a diesel (CI) engine. Since the calorific value of standard diesel fuel ranges between 43–47 MJ/kg, so the diesel used for the comparative study had a value of 46.6±0.2 MJ/kg which was noted to be higher than the blended fuel. But the calorific value of the blended fuel (43.98±0.5 MJ/kg) was within the range of the calorific value of the standard diesel. Since the properties of the model-optimized blend fuel complied with the specified range of diesel standards and were much comparable to the diesel fuel. Accordingly, less impact on the operating characteristics of the engine is expected from the substitution of diesel with the blended fuel in a conventional CI engine. However, lower carbon content was noted for the model-optimized blend as compared to diesel. The lower carbon content was noted because of increased aliphatic content 65.8% and reduced aromatic content 8.4% as compared to diesel having 30.6% aliphatic content and 25.8% aromatic content. Thus, the model optimized blend can be expected to have a cleaner emission performance than diesel.

Table 3.9 Comparison of fuel properties of the model-optimized blend with diesel fuel.

Fuel properties	Carbon content (%)	Density at 15 °C (g/ml)	Viscosity at 40 °C (mm²/s)	Flash point (°C)	Fire point (°C)	Calorific value (MJ/kg)	Cetane index
Model-optimized blend	81.8	0.829±0.004	4.36±0.12	63±2	71±4	43.98±0.5	49±1
Diesel fuel	86.6	0.823±0.003	3.38±0.08	48±3	58±5	46.65±0.2	52±1
Test method		ASTM D1217	ASTM D2893	ASTM D93	ASTM D93	ASTM D240	ASTM D976
IS 1460 Diesel standard		0.810-0.845	2.0-4.5	> 35	----	> 43.0	> 46

The performance and emission profiles of the model-optimized blend were assessed in a conventional CI engine under two different engine load conditions. High engine load conditions corresponding to 80% of full load and low engine load conditions corresponding to 20% of full load, were the two different engine load conditions evaluated for the fuels. The BSFC, BTE, ME, and VE were the performance parameter assessed for the blended fuel. The performance parameters of the blend were compared against diesel fuel. The comparison of the performance and emission characteristics of the model-optimized blend against diesel is tabulated in Table 3.10.

BSFC, a measure of the mass of fuel required to produce the desired output power in an engine, was noted to be significantly lower for the model-optimized blend compared to diesel. But, BTE, defined as the measure of the efficiency to convert the chemical energy of a fuel into useful mechanical work upon combustion, was recorded as significantly higher for blend in comparison to diesel. Literature suggests that lower viscosity favors better atomization and spray characteristics and increased volatility causes faster mixing and better combustion [9]. Also, the presence of oxygenated fuel (like WVO-BD) in a blend fuel promotes better combustion and reduced frictional loss [25]. Thus, comparable density and lower viscosity of the blend may be attributed to better fuel economy (lower BSFC) and improved thermal efficiency (higher BTE) relative to diesel.

The ME, the ability of the engine to translate the indicated power produced from fuel combustion into the brake power, was found to be comparable for the blend and diesel for all engine load conditions. A comparable ME between model-optimized blend and diesel confirmed that substitution of diesel with the blended fuel did not result in additional losses in the engine. The VE, a ratio of the air consumed during combustion to the cylinder capacity, is known to influence the engine output. VE of the model-optimized blend was also noted to be comparable with diesel for all engine load conditions. A comparable VE between model-optimized blend and diesel concluded that blended fuel has combustion efficiency analogous to diesel. Thus, the analysis of performance parameters established the model-optimized blend as a potential diesel alternative.

Table 3.10 Comparison of engine performance and emissions of the model-optimized blend with diesel fuel.

Engine performance characteristics								
	BSFC (kg/kW-h)		BTE (%)		ME (%)		VE (%)	
Engine load Fuels	Low	High	Low	High	Low	High	Low	High
Model-optimized blend	0.629 ±0.024	0.252 ±0.017	13.13 ±0.79	31.41 ±0.67	73.65 ±0.60	80.64 ±0.50	86.54 ±0.50	77.50 ±0.70
Diesel fuel	0.643 ±0.028	0.262 ±0.017	11.98 ±0.69	29.42 ±0.63	73.42 ±0.51	79.79 ±0.30	86.83 ±0.60	77.85 ±0.60
Engine emission characteristics								
	CO (g/kWh)		CO_2 (g/kWh)		HC (g/kWh)		NO_x (g/kWh)	
Engine load Fuels	Low	High	Low	High	Low	High	Low	High
Model-optimized blend	6.448 ±0.14	0.772 ±0.09	315.82 ±10.21	272.99 ±9.28	0.414 ±0.003	0.332 ±0.003	3.908 ±0.09	5.085 ±0.11
Diesel fuel	7.591 ±0.15	1.159 ±0.12	208.95 ±9.52	209.52 ±10.12	0.376 ±0.002	0.291 ±0.003	3.724 ±0.08	4.995 ±0.12

The emission characteristics of the model-optimized blend fuel were compared against diesel at high and low engine load conditions. The comparative emission characteristics are summarized in Table 7. The levels of carbon monoxide (CO) in the emission is a measure of the incomplete combustion whereas carbon dioxide (CO_2)

emitted from the engine shows the completeness of the fuel combustion. The presence of oxygenated fuel like WVO-BD increased the oxygen content in the blended fuel resulting in more completeness of combustion. As a result, lower CO emission and higher CO_2 emission were recorded for the blended fuel as compared to diesel during combustion in the CI engine. The emissions of oxides of nitrogen (NO_x) and unburnt hydrocarbon (HC) of the optimized blend were found to be very comparable with that of diesel. Thus, substitution with the model-optimized blend fuel was identified as a cleaner alternative to diesel fuel for utilization in a CI engine.

3.4 Summary

The study presented a value-addition route for biodiesel prepared from WVO through utilization as a blend in a diesel engine. Since density and calorific value are the prime fuel properties that govern the performance of a fuel in engine combustion. So, the density and calorific value of the WVO-BD blend were modulated by varying the WVO-BD fraction in the blend, density of other fuel oil used in the blend, and using different WVO-BD stocks prepared from WVO with varied thermal exposure. The density of the blend was modeled using the RS approach to define an optimum blend fuel that is equivalent to diesel fuel. The factors considered for predicting the density of the blend were: WVO-BD fraction in the blend, thermal exposure time of WVO (h), and fuel oil density (g/ml). The RS model was assessed for prediction accuracy and validated against additional blends prepared experimentally. The results established that the model can accurately predict (with R^2 of 0.904) for different factors at all the factor-level considered in this investigation. The optimum factor setting for a blend with a density equivalent to standard diesel was predicted using the RS model. A blend density of 0.827 g/ml was predicted for an optimum blend of WVO-BD prepared from WVO with 48 hours of thermal exposure with fuel oil of density 0.80 g/ml at a blend fraction of 0.46. An equivalent blend when prepared experimentally was noted to have a density of 0.829±0.004 g/ml. The error of prediction was computed to be less than 1.0% between the density model-optimized blend and equivalent experimental blend. The blend density was used to predict the calorific values of the blended fuel using established relations. The model-based prediction of calorific values was validated for different experimental factors over entire factor-level settings. The model was found to predict the calorific value of the blend with 99% accuracy. The calorific value of 44.5 MJ/kg for the optimum blend was found to comply with the calorific values of the standard diesel. The fuel properties, engine performance, and emission characteristics of the model-optimized blend were compared against stand diesel. The properties of the blended fuel complied with the specified range of diesel standards and were much comparable to the diesel. A comparison of engine performance parameters established the model-optimized blend as a potential substitute for diesel fuel in CI engines. Also, model-optimized blend fuel was identified as a cleaner alternative to diesel fuel through emission analysis. A reduction of 43 - 46% in engine operating cost is achievable with the model-optimized blend relative to diesel. Thus, the study presented a value-addition route of WVO-BD and defined a real-life WVO-BD-based diesel-substitute fuel that can have commercial, societal, environmental, and financial benefits.

3.5 References

[1] Suresh, M., Jawahar, C. P., Richard, A. (2018), "A review on biodiesel production, combustion, performance, and emission characteristics of non-edible oils in variable compression ratio diesel engine using biodiesel and its blends", Renew Sustain Energy Rev, v. 92, pp. 38-49.

[2] Kathirvel, S., Layek, A., Muthuraman, S. (2016), "Exploration of waste cooking oil methyl esters (WCOME) as fuel in compression ignition engines: a critical review", Eng Sci Technol Int J, v. 19, pp. 1018-1026.

[3] Borugadda, V. B., Paul, A. K., Chaudhari, A. J., Kulkarni, V., Sahoo, N., Goud, V. V. (2018), "Influence of Waste Cooking Oil Methyl Ester Biodiesel Blends on the Performance and Emissions of a Diesel Engine", Waste Biomass Valor, v. 9, pp. 283-295.

[4] Wei, L., Cheng, R., Mao, H., Geng, P., Zhang, Y., You, K. (2018), "Combustion process and NOx emissions of a marine auxiliary diesel engine fueled with waste cooking oil biodiesel blends", Energy, v. 144, pp. 73-80.

[5] Attia, A. M. A., Hassaneen, A. E. (2016), "Influence of diesel fuel blended with biodiesel produced from waste cooking oil on diesel engine performance", Fuel, v. 167, pp. 316-328.

[6] Yesilyurt, M. K. (2019), "The effects of the fuel injection pressure on the performance and emission characteristics of a diesel engine fueled with waste cooking oil biodiesel-diesel blends", Renew Energy, v. 132, pp. 649-666.

[7] Prabu, S. S., Asokan, M. A., Roy, R., Francis, S., Sreelekh, M. K. (2017), "Performance, Combustion and Emission Characteristics of Diesel Engine fueled with Waste Cooking Oil Bio-diesel or diesel blends with Additives", Energy, v. 122, pp. 638-648.

[8] Labeckas, G., Slavinskas, S. (2015), "Combustion phenomenon, performance and emissions of a diesel engine with aviation turbine JP-8 fuel and rapeseed biodiesel blends", Energy Convers. Manag, v. 105, pp. 216–229.

[9] Aydın H (2016), "Scrutinizing the combustion, performance and emissions of safflower biodiesel–kerosene fueled diesel engine used as power source for a generator", Energy Convers Manag, v. 117, pp. 400-409.

[10] Bayindir, H., Isık, M. K., Argunhan, Z., Yücel, H. L., Aydın, H. (2017), "Combustion, performance and emissions of a diesel power generator fueled with biodiesel-kerosene and biodiesel-kerosene-diesel blends", Energy, v. 123, pp. 241-251.

[11] Chen, H., He, J., Chen, Y., Hua, H. (2018), "Performance of a common rail diesel engine using biodiesel of waste cooking oil and gasoline blend", J Energy Institute, v. 91, pp. 856-866.

[12] Putrasari, Y., Lim, O. (2019), "A Review of Gasoline Compression Ignition: A Promising Technology Potentially Fueled with Mixtures of Gasoline and Biodiesel to Meet Future Engine Efficiency and Emission Targets", Energies, v. 12, pp. 238.

[13] Gulum, M., Bilgin, A. (2015), "Density, flash point and heating value variations of corn oil biodiesel–diesel fuel blends", Fuel Process Technol, v. 134, pp. 456-464.

[14] Ramírez-Verduzco, L. F., Rodríguez-Rodríguez, J. E., & del Rayo Jaramillo-Jacob, A. (2012), "Predicting cetane number, kinematic viscosity, density and higher heating value of biodiesel from its fatty acid methyl ester composition". Fuel, v. 91(1), pp. 102-111.

[15] Myers, R. H., Montgomery, D. C., Anderson-Cook, C. M. (2016), Response Surface Methodology: Process and Product Optimization using Designed Experiments, 4th Edition, John Wiley & Sons, NJ.

[16] Majhi, S., Ray, S. (2016), "A study on production of biodiesel using a novel solid oxide catalyst derived from waste", Environ Sci Pollut Res, v. 23, pp. 9251-9259.

[17] Sharoba, A. M., Ramadan, M. F. (2012), "Impact of Frying on Fatty Acid Profile and Rheological Behaviour of Some Vegetable Oils", J Food Process Technol, v. 1, pp. 3-7.

[18] Kumar, N. (2017), "Oxidative stability of biodiesel: Causes, effects and prevention", Fuel, v. 190, pp. 328-350.

[19] Vempatapu, B. P., Kanaujia, P. K. (2017), "Monitoring petroleum fuel adulteration: A review of analytical methods", Trends Analytical Chemistry, v. 92, pp. 1-11.

[20] Box, G. E.P., Draper, N. R. (1987), Empirical Model Building and Response Surfaces. John Wiley and Sons, NY, pp. 205–477.

[21] Demirbas, A. (2006), "Theoretical Heating Values and Impacts of Pure Compounds and Fuels", Energy Sources Part A Recov Util Environ Effects, v. 28, pp. 459-467.

[22] Demirbas, A. (2008), "Relationships derived from physical properties of vegetable oil and biodiesel fuels", Fuel, v. 87, pp. 1743-1748.

[23] Tripathy, D. B., Mishra, A. (2017), "Microwave Synthesis and Characterization of Waste Soybean Oil-Based Gemini Imidazolinium Surfactants with Carbonate Linkage". Surface Rev Letters, v. 24(05), pp. 1750062.

[24] Uğuz, G., Atabani, A. E., Mohammed, M. N., Shobana, S., Uğuz, S., Kumar, G., Ala'a, H. (2019), "Fuel stability of biodiesel from waste cooking oil: A comparative evaluation with various antioxidants using FT-IR and DSC techniques". Biocatalysis Agricultural Biotechnol, v. 21, pp. 101283.

[25] Agarwal, A. K., Dhar, A. (2013), "Experimental investigations of performance, emission and combustion characteristics of Karanja oil blends fuelled DICI engine", Renew. Energy, v. 52, pp. 283–291.

Chapter 4

TO COMPARATIVELY ANALYZE WASTE VEGETABLE OIL AND WASTE VEGETABLE OIL-BIODIESEL IN DIESEL AS ALTERNATIVE FUEL FOR A CI ENGINE

4.1 Introduction

Reports are available with regards to the evaluation of the transesterified waste vegetable oil or WVO-biodiesel (WVO-BD) as an alternative fuel fraction in diesel blend for CI engines. The lower calorific value of WVO-BD blends result in lower torque generation but produce cleaner emissions than diesel fuel. Blends containing up to 0.50 WVO-BD fraction were reported to perform well in CI engines. Also, the fuel injection pressure between 190-210 bars was reported to provide better engine performance results for diesel blends with high biodiesel fraction [1-2].

The utilization of WVO (without transesterification) in diesel blends is rather less explored. A couple of studies are present in the literature in which WVO was preheated before feeding it to the CI engine along with diesel [3]. Preheating lowers the viscosity and improves the spray property of the fuel. Because a preheated WVO has viscosity 10 times lower than its viscosity at room temperature. A fuel with lower viscosity has better spray characteristics than fuel with high viscosity. From the literature survey (presented in Chapter 2), it has been established that utilization of WVO, directly or after transesterification, should prove to be advantageous in managing a common liquid waste and also providing alternative fuel fraction. A prominent research question has arisen – is transesterification of WVO is required to valorize it as an alternative fuel fraction in diesel? The answer to this question is attempted here.

Accordingly, the objective of the study is to evaluate WVO against WVO-BD in diesel blends for engine performance parameters and emission characteristics in a research CI engine. An economic assessment will be made to understand the monetary incentive linked with the utilization of these blend fuels as a diesel replacement in a CI engine. A thorough literature search has confirmed that a similar comparative assessment, to choose the efficient and economic blend for diesel substitution was never being framed earlier, and similarly, a comparative assessment between WVO and WVO-BD was also not reported.

4.2 Methodology

4.2.1 Materials

Diesel was sourced from a local diesel fueling station. WVO and WVO-BD were prepared as per the methodology described in Section 3.2.2 in Chapter 3.

4.2.2 Preparation of diesel blends

The volumetric blend of WVO with diesel was prepared in three different blend proportions, namely, 0.20, 0.35, and 0.50 fractions as WVO. Similarly, WVO-BD-diesel blends were also prepared by volumetric blending with diesel in the blend fraction of 0.20, 0.35, and 0.50 of WVO-BD. Blend fraction above 0.50 was not considered as conventional diesel engines have not been reported to accommodate higher WVO or WVO-BD fraction in blend fuel.

Thus, the different blends evaluated in the study were: WVO20 (0.80 fraction diesel + 0.20 fraction WVO), WVO35 (0.65 fraction diesel + 0.35 fraction WVO), WVO50 (0.50 fraction diesel + 0.50 fraction WVO), WVO-BD20 (0.80 fraction diesel + 0.20 fraction WVO-BD), WVO-BD35 (0.65 fraction diesel + 0.35 fraction WVO-BD), and WVO-BD50 (0.50 fraction diesel + 0.50 fraction WVO-BD).

4.2.3 Characterization of test fuels

The WVO and WVO-BD were characterized using ASTM standard test methods (as described in section 3.2.3) and compared with ASTM D975/ IS 1460 diesel standards and ASTM D6751 biodiesel standards, respectively. The fatty acid profiles of WVO and WVO-BD were determined using GC-MS analysis, as mentioned in Section 3.2.3 in Chapter 3.

4.2.4 Assessment of engine performance and emission characteristics

The test engine setup was the same as employed for the study presented in Section 3.2.4 in Chapter 3. The accuracy of the results obtained was defined from the standard deviation associated with the measurement. All measurements were made in triplicate to determine the standard deviation. The uncertainties associated with various measurements were calculated by the method reported in the literature [4]. The overall uncertainty of the measurements was computed as less than 5.0 %.

4.3 Experimental findings

4.3.1 Assessment of fuel properties

The WVO reportedly comprised of 22.7 % (w/w) saturated fatty acids (SFA) (C16:0 and C18:0 fatty acids), 35.9 % (w/w) mono-unsaturated fatty acid (MUFA) (C18:1 fatty acid), and 40.8 % (w/w) poly-unsaturated fatty acids (PUFA) (C18:2 and C18:3 fatty acids), whereas the free fatty acid content was 1.04 % (w/w). Upon transesterification, the fatty acid composition of WVO-BD comprised 24.1 % (w/w) SFA, 34.2 % (w/w) MUFA, and 40.2 % (w/w) PUFA, with no free fatty acid in it.

The operating characteristics, such as ignition delay, the lubricity of fuel, fuel economy, and available power greatly influence the engine's performance. Literature suggests that for a particular engine type and design, these operating parameters are thus affected by fuel properties [5]. Thus, the determination of fuel properties is important for understanding and estimating the engine performance of the fuel. As per ASTM standard test methods, the blend fuels were characterized in terms of density, viscosity, flash point, calorific value, and cetane index. The fuel properties of the afore-mentioned fuels were measured and tabulated in Table 4.1.

Table 4.1 The measured properties of different fuel stocks.

Fuel Properties	ASTM Test Methods	Diesel	WVO	WVO-BD	Standard Diesel (ASTM D975)	Standard Bio-diesel (ASTM D6751)
Carbon content (wt%)		86.6	76.4	76.9	----	----
Specific gravity at 15 °C (kg/m^3)	D 1217	0.823 ±0.003	0.893 ±0.002	0.882 ±0.003	0.82-0.845	0.80 - 0.87
Kinematic viscosity at 40 °C (mm^2/s)	D 2983	2.612 ± 0.08	31.580 ±0.04	3.780 ±0.07	1.9-4.1	1.9 - 6.0
Flash point (°C)	D 93	48±2	> 200	135±4	51 min.	92 min.
Cetane index	D4737/ D 976	51.25 ±0.58	48.28 ±0.81	55.32 ±0.76	40 min.	46 min.
Gross calorific value (MJ/kg)	D 240	46.65 ±0.12	39.90 ±0.31	41.82 ±0.28	43.00 min.	----

Note: min. – minimum value.

The density of the fuel is an important property to define the evaporation characteristics of fuel which affect the combustion and economy of the fuel. Higher density causes higher mass injection for the same volume of fuel injected, resulting in lower fuel economy. The densities of the blends (Table 4.2) were proportional to the blending fraction of WVO and WVO-BD. Accordingly, the densities of WVO blends were higher than equivalent WVO-BD blends. The densities of all the D-WVO and D-WVO-BD blends were within the specified limit for the density range (0.820 - 0.860 g/ml) of standard diesel. Compliance of the blend density with the specification range ensured that engine power will not be significantly altered by the replacement of the diesel with the blend fuels.

Flashpoint is a very important property related to the combustion and transportation of fuel. A higher flash point (or lower volatility) influences the evaporation characteristics but makes the fuel safer for handling and storage. An increased fraction of diesel in the D-WVO and D-WVO-BD blends reduced the flashpoint of the blended fuel (Table 4.2). The flashpoints of all WVO and WVO-BD blend fuels were higher than that of diesel but comply with the diesel standards. Thus safer handling and ease of storage and transportation are expected of all the blend fuels.

The kinematic viscosity plays an important role in the atomization and lubrication property of the fuel. Although higher viscosity minimizes jet penetration but favors a reduction in fuel loss causing faster evolution of pressure inside the cylinder. The WVO had higher kinematic viscosity as compared to WVO-BD and the kinematic viscosity of WVO-BD was higher than that of diesel (Table 4.2). The viscosity of the WVO-BD used in the study complied with the bio-diesel standard. An increase in diesel fraction in the blends reduced the viscosity for either blend type (Table 4.1). Higher viscosity was recorded for all D-WVO blends as compared to D-WVO-BD blends. The viscosity of all WVO-BD-diesel complied with the viscosity range of diesel standard but the viscosity of the WVO-diesel blends exceeded the standard range (1.9 - 4.1 mm^2/s) for diesel. So, somewhat lower fuel consumption should be expected for the D-WVO-BD blends than their WVO counterparts.

The cetane index (cetane number equivalence) is an estimator for the ignition delay (or readiness to spontaneous combustion) of the fuel in the cylinder of a CI engine.

Higher cetane index (or cetane number) ensure shorter ignition delay and better ignition quality resulting in smoother engine operation. The cetane index of WVO was lower than diesel and that of diesel was lower than WVO-BD (Table 4.1). Accordingly, the cetane index of the blends varied proportionately with the change in WVO or WVO-BD fraction. Overall, the cetane indices of the WVO blends were marginally lower than WVO-BD blends (Table 4.2). The cetane index of the D-WVO-BD varied between 48-56 and those of D-WVO blends between 48-55. Analogous cetane indices show that not much difference between the ignition delays should be expected for D-WVO and D-WVO-BD blends. Thus, minimum variation in ignition delay can be expected due to a change in fuel for the CI engine during the study.

The calorific value is a measure of the energy available from fuel and is linked to the economy of the fuel. The calorific values of WVO and WVO-BD were lower than diesel (Table 4.1). Therefore, the calorific values of D-WVO blends and D-WVO-BD blends showed an increasing trend with an increase in the diesel fractions in the blend. However, the calorific values of all the blend fuels were comparable with diesel and complied with the diesel standard. Among the blends, the calorific values of D-WVO blends were quite akin to those of D-WVO-BD blends, for a fixed diesel fraction. This shows that somewhat similar power is expected to be obtained from the combustion of both types of diesel blends inside a CI engine.

Table 4.2 The measured properties of WVO and WVO-BD blends with diesel.

Fuel blends Property	D100	WVO20	WVO35	WVO50	WVO-BD20	WVO-BD35	WVO-BD50
Specific gravity at 15 °C (kg/m³)	0.823 ±0.003	0.841 ±0.002	0.851 ±0.003	0.860 ±0.002	0.836 ±0.004	0.844 ±0.003	0.855 ±0.002
Kinematic viscosity at 40 °C (mm²/s)	2.612 ±0.08	4.224 ±0.12	6.328 ±0.18	9.464 ±0.19	3.101 ±0.22	3.264 ±0.25	3.466 ±0.18
Flash point (°C)	48 ±2	48 ±3	49 ±3	55±2	48 ±2	51 ±2	56 ±3
Cetane index	51.25 ±0.58	50.87 ±0.61	50.16 ±0.54	49.67 ±0.75	51.96 ±0.38	52.63 ±0.65	53.29 ±0.26
Calorific value (MJ/kg)	46.65 ±0.12	45.32 ±0.11	44.37 ±0.18	43.80 ±0.17	45.55 ±0.20	44.82 ±0.27	43.92 ±0.21

Except for the viscosity of the WVO-blends, the properties of the blend fuels complied with the specified range of diesel standards and were much comparable to the diesel. Accordingly, less impact on the operating characteristics of the engine is expected from the fuel change over the fuel change over on the CI engine at a particular load setting.

4.3.2 Assessment of performance characteristics of blend fuels

The engine performance assessment indicates the extent of conversion of the chemical energy of the fuel into useful mechanical work in the engine. Accordingly, mechanical efficiency (ME), volumetric efficiency (VE), brake specific fuel consumption (BSFC), and brake thermal efficiency (BTE) were assessed.

The mechanical efficiency (ME) is the ratio of the measured performance to the performance of an ideal engine. The ME does not depend on fuel consumption, rather it is the ratio of the brake power developed due to engine load to the indicated power produced from the combustion of fuel inside the cylinder. As evident from Fig. 4.1(a), the ME was equal for all the test fuels at lower load conditions. As all the blend fuels have produced similar or close brake power for input power, so consequently the ME was found to be nearly the same for all the blends. Inherent lubricity of WVO and WVO-BD may be attributed to lower frictional loss resulting in marginally better ME for the blend fuels relative to diesel at higher load settings. Similar results are reported elsewhere [6, 7]. However, the ME was largely comparable for all the test fuels. The observation is in agreement with the trend of the properties of the test fuels, wherein properties of most fuel blends were noted to be comparable to diesel.

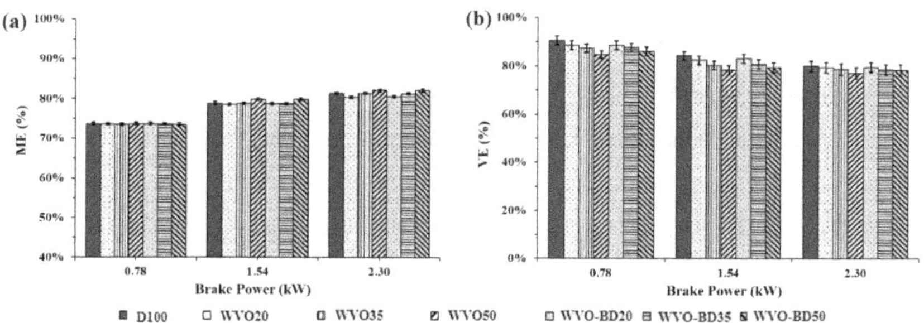

Fig. 4.1 Comparison of (a) ME and (b) VE.

The volumetric efficiency (VE) corresponds to the ratio of the amount of air consumed during the combustion of fuel to the displacement volume of the engine cylinder. Thus, the higher the VE better the combustion. The variations in the VE of the blend fuels relative to diesel for different load conditions are presented in Fig. 4.1(b). Increased engine load requires more fuel to provide the desired power which limits the air for combustion resulting in a decreased VE. Lower VE has been reported to cause a rise in the cylinder temperature which affects the emission properties [8, 9]. VE for D-WVO blends and D-WVO-BD blends were noted to trail behind that for diesel. Lower VE for the blends may be attributed to a slightly higher density of the blend fuels. Also, the VE of WVO-BD blends was observed to be marginally better than WVO blends. The oxygen-rich composition of blend fuels might also be responsible for the increase in combustion temperature and thereby limiting the VE of the blends relative to diesel. A comparable VE of the fuels can be concluded to have analogous combustion efficiency, particularly at high load (brake power) settings.

The brake-specific fuel consumption (BSFC) is the rate at which the fuel is injected into the engine for producing the desired power at a specific engine load. A comparison of various blends against diesel for different engine loads is presented in Fig. 4.2(a). BSFC was noted to reduce with the increase in engine load or brake power. Improved combustion efficiency at higher engine loads shall be attributed to a reduced fuel consumption [8]. The comparison demonstrates that the BSFCs of WVO blends were somewhat higher than that of WVO-BD blends. The higher densities of D-WVO blends caused higher mass injection for the same volume of fuel leading to an increased BSFC. The WVO50 blend showed an 11.7% and 14.6% higher BSFC than diesel and WVO-BD50, respectively at low load conditions. Under low brake power, the BSFCs of blends with lower WVO-BD fractions were noted to be slightly better than diesel. Comparable fuel properties but increased oxygen content have resulted in better combustion of these blends relative to diesel. Higher engine load conditions caused an increase in the in-cylinder temperature resulting in better fuel mixing and combustion of the fuels. Accordingly, all the blends have BSFCs comparable with diesel at higher brake power (load).

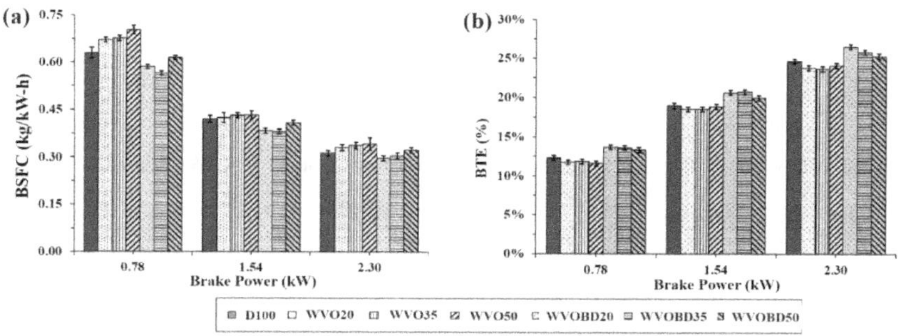

Fig. 4.2 Comparison of **(a)** BSFC and **(b)** BTE.

The brake thermal efficiency (BTE) is a measure of the efficiency of an engine to convert the chemical energy in a fuel into mechanical work done upon combustion. The comparison of the BTE of various D-WVO and D-WVO-BD blends against diesel is shown in Fig. 4.2(b). BTE was noted to increase with an increase in engine load. Lower BSFC at higher engine load (brake power) may be attributed to an enhanced BTE [8]. Under all load conditions, the BTE of all the D-WVO blends were marginally lower than diesel whereas the D-WVO-BD blends showed consistently higher BTE compared to diesel or WVO blends. The diesel-like properties with higher oxygen content might be responsible for improved BTE of WVO-BD blends compared to diesel. Higher oxygen content with equivalent viscosity enhances the oxidation during combustion but minimizes frictional loss resulting in better BTE for the WVO-BD blends. Thus, for higher blend fraction BTE decreases due to reduced calorific value and increased density. Increased BTE for higher blend fraction agreed with the trend of BSFC. A maximum difference of 7.7% in BTE was recorded for WVO-BD20 relative to diesel at the low engine load condition. Largely, the BTEs of all the fuels were comparable at high brake power (load) settings.

4.3.3 Assessment of emission characteristics of blend fuels

The combustion condition in the cylinder of an engine has been reported to depend on the fuel properties and performance characteristics of the fuel during engine operation. Accordingly, an assessment of the emission characteristics of the various test fuels under different load conditions holds pertinence in the study. The emission characteristics were assessed in terms of specific emissions of carbon monoxide (CO), carbon dioxide (CO_2), unburnt hydrocarbon (HC), and oxides of nitrogen (NO_x). The emission characteristics of D-WVO and D-WVO-BD blends were determined and compared with diesel fuel at different engine load conditions. Although particulate emission is a major problem of CI engines and vegetable oil-based blends are known to produce particulates. However, due to technical limitations, the same cannot be assessed, which is a limitation of the study.

Carbon monoxide (CO) is emitted from the exhaust of CI engines due to the incomplete combustion of fuel inside the cylinder. The trend for carbon CO emission for various blends in comparison to diesel is presented in Fig. 4.3(a). The CO emission was noted to decrease with the increase in the engine load. The temperature of the gas inside the engine cylinder remains low under low load conditions resulting in incomplete combustion and relatively high emission of CO. Increase in the combustion temperature at higher engine loads favors complete combustion of fuel and lowers CO emission. Reduced CO emission was recorded for the blended fuels in comparison to diesel at all load conditions, particularly at low load conditions (brake power). At the high load condition, the CO emissions of the blends were closely comparable to diesel. Also, the increase in WVO and WVO-BD fractions in the blend caused a decrease in the CO emission. The highest CO reduction (-50.6% relative to diesel) was recorded for the WVO-BD50 blend as compared to the WVO50 blend (-23% relative to diesel) at 0.78 kW load. WVO and WVO-BD have higher oxygen content (~ 10% more) as compared to diesel. The higher oxygen content in the blend may be accounted for better oxidation during combustion resulting in lesser CO emission. The WVO contains close to 12% oxygen by mass. While diesel has no oxygen molecule in it. As a result, the D-WVO blend has higher oxygen content than diesel. The fuel-bound oxygen facilitates the combustion of the fuel inside the engine cylinder. As a result, the fuel combustion nears completeness resulting in lowered CO formation. The results obtained for CO emission were in agreement with the results reported in the literature for biodiesel blends [9, 10].

The amount of carbon dioxide (CO_2) in the exhaust indicates the completeness of the combustion of fuel inside the engine cylinder. A comparison of CO_2 emission for various blend fuels against diesel is presented in Fig. 4.3(b). A significantly higher CO_2 emission was noted for all the blended fuels as compared to diesel under all load conditions. Better combustion for oxygenated fuel, like blend fuel, was reported due to higher combustion temperature [9]. Completeness of combustion can be attributed to better conversion of CO to CO_2 and increased CO_2 emission for blend fuels. The same was confirmed using a mass balance. The mass balance showed that at a particular engine load, the % mass reduction of CO was directly proportional to the % mass elevation on CO_2 in the exhaust gases. Accordingly, increased WVO or WVO-BD fraction in the blend resulted in higher CO_2 emission than diesel. A comparable CO_2 emission was for all WVO blends with respect to the WVO-BD blends. Decreased CO_2 emission at increased load (brake power) setting for all the fuels may be credited to a lower BSFC under increased load conditions (Fig. 4.2(a)) for all the test fuels. Although higher CO_2 amounts were emitted by the blended fuels, the use of WVO blends or WVO-BD blends as fuel favors better life-cycle circulation of CO_2 due to renewable origin.

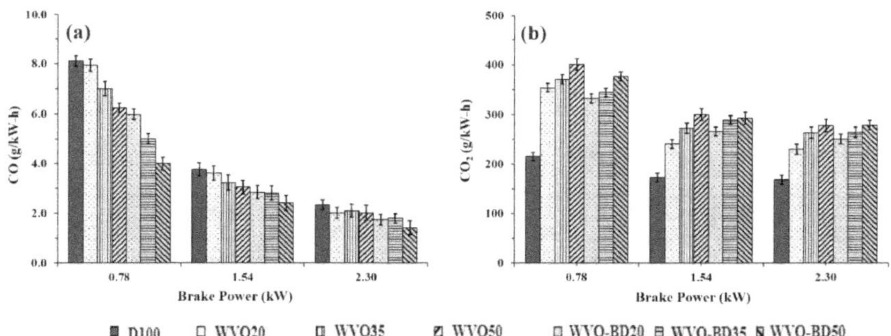

Fig. 4.3 Comparison of **(a)** CO and **(b)** CO_2 emission.

The unburnt hydrocarbon (HC) emission from the engine takes place due to the incomplete combustion of fuel. The specific emissions of HC of all the test fuels for different load conditions are presented in Fig. 4.4(a). Literature reported higher HC emission at low cylinder temperature under the lower load setting due to a reduced rate of oxidation [10]. Thus, specific emissions of HC of all the test fuels were observed to decrease with the increase in engine load. An increase in the blend fraction of WVO or WVO-BD was observed to have a marginal effect on HC emission. Increased WVO or WVO-BD fraction raised the viscosity and oxygen content of the blend. Increased oxygen content was known to promote better combustion, whereas increased viscosity affects the atomization of fuel that hinders fuel combustion [11]. Due to these contradictory effects, only a marginal change in HC emission was noted for the change in blend composition. Expect at lower load conditions, the HC emissions of WVO-diesel blends were found to be comparable with WVO-BD blends. Increased viscosity of WVO blends together with less cylinder temperature may be accounted for higher HC emission of the WVO blends at low load conditions. Overall, the fuels have a largely indistinguishable level of HC emission at a higher load (brake power) setting.

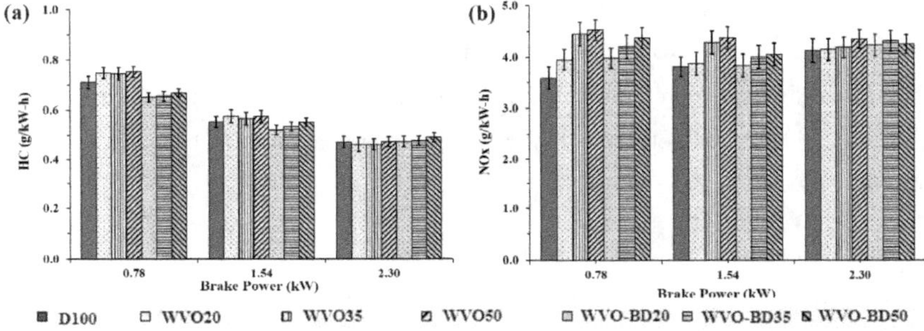

Fig. 4.4 Comparison of **(a)** HC and **(b)** NOx emission.

The nitrogen of the intake air in the combustion cylinder of the engine is converted to oxides of nitrogen (NO_x) under combustion conditions where in-cylinder temperatures tend to become quite high. Since NO_x formation has high activation energy so high combustion temperature favors NO_x formation. The NO_x emission of blend fuels and diesel under various load (brake power) settings is presented in Fig. 4.4(b). Incremental NO_x emission was noted for increased load settings for all test fuels. Increased load causes a rise in the cylinder temperature which can be accounted for by the observation. An increase in WVO or WVO-BD fraction in the blend was noted to increase the NO_x emission. The higher proportion of oxygenated fuel (*i.e.* WVO or WVO-BD) in the blend increased the in-cylinder temperature that favored higher NO_x emission. Under all load conditions, the NO_x emission of WVO blends and WVO-BD blends were found to be comparable. Largely, the NO_x emissions of all test fuels were nearly equivalent, varied between 3 – 5%, at the high load condition.

4.3.4 Assessment of the economics for the blend fuels

The comparative price summary of various D-WVO blends and D-WVO-BD blends against diesel fuel is presented in Table 4.3. The reported price for a liter of WVO was $0.21 [12], WVO-BD was $0.61 [13] and diesel was $0.95 [14], respectively. The price per liter of the blend was computed from the reported fuel price. The fuel cost was calculated from the fuel consumption rate and fuel price. The percent reduction in fuel cost was determined against the fuel cost with diesel as reference. The overall price of biodiesel depends upon the cost of feedstock and the cost of transesterification. Even if the low-cost feedstock is considered, the cost of transesterification makes D-WVO-BD blends costlier than D-WVO blends. Thus, a comparison between WVO-diesel blends and D-WVO-BD blends with reference to diesel is necessary to establish the choice of the economic blend for the assessment of sustainability. WVO20 and WVO-BD20 blends showed a comparable reduction in fuel cost for running the test engine for an hour. But, increased WVO fraction in the blend significantly reduced the engine running cost as compared to the equivalent D-WVO-BD blend. WVO50 blend was computed to be more economic (being 35.8% more economic than diesel) compared to the WVO-BD50 blend (which showed 18.53% lower fuel cost reduction compared to diesel).

Table 4.3 Economic comparison of WVO and WVO-BD blends against diesel.

Fuel and Blends	Diesel (Cost per Liter)	WVO and blends (Cost per Liter)	Engine running cost per hour[#]	% cost reduction for D-WVO blends	WVO-BD and blends (Cost per Liter)	Engine running cost per hour[#]	% cost reduction for D-WVO-BD blends
W100*	---	$ 0.210	---	---	$ 0.609	---	---
D100	$ 0.950	---	$ 0.831	---	--	---	---
W*20	---	$ 0.802	$ 0.726	12.64 %	$ 0.882	$ 0.723	12.99 %
W*35	---	$ 0.691	$ 0.631	24.07 %	$ 0.831	$ 0.693	16.60 %
W*50	---	$ 0.580	$ 0.532	35.98 %	$ 0.779	$ 0.677	18.53 %

Note: W* - denotes either WVO or WVO-BD; [#] - considered at the highest engine load setting.

4.4 Summary

The study presented a comparative assessment of WVO and WVO-BD in diesel blends as an alternative renewable fuel for CI engines. WVO and WVBD were analyzed in different blend fractions with diesel. WVO-diesel and WVO-BD-diesel blends were noted to comply with the diesel standards for density, flash point, cetane indices, and calorific values. The viscosities of WVO-diesel blends were found to be higher than the diesel standard. But, the viscosities of WVO-BD-diesel blends were within the standard range for diesel. ME and VE of all the diesel blends were very much comparable and nearly equivalent to diesel at the higher engine load condition. BSFC of the WVO-diesel blends were higher, while BSFCs of WVO-BD-diesel blends were lower than diesel at the high engine load condition. WVO-BD-diesel blends showed a 2-5 % reduction in BSFC when compared to diesel fuel. BTE of all the WVO-diesel blends were comparable with diesel, while the WVO-BD-diesel blends showed consistently higher BTE than diesel. Higher oxygen content in WVO and WVO-BD resulted in lower CO emission and higher CO_2 emission for WVO-diesel and WVO-BD-diesel blends as compared to diesel. Slightly lower HC emission and marginally higher NO_x emissions were noted for all WVO-diesel and WVO-BD-diesel blends with reference to diesel. Fuel cost analysis established WVO-diesel blends as a more economical alternate to diesel than WVO-BD-diesel blends. 35.8% saving in fuel cost per hour relative to diesel was established for the WVO50 blend as compared to 18.53% computed for the equivalent WVO-BD50 blend.

The study established that blended fuels of WVO-BD or WVO are both sustainable and economic with comparable engine performance and better emissions with regard to diesel. But the WVO50 blend is more sustainable and economical against the predominant and more explored WVO-BD blend as the diesel fuel alternative for CI engines. The study establishes a paradigm shift in the utilization scope for WVO wherein WVO can be directly used without transesterification to WVO-BD as a blended fuel in a diesel engine without hampering the engine performance. The study also presents a valorization route for voluminous liquid waste, WVO, towards sustainable energy solutions. An absence of a similar comparative analysis of WVO-BD against WVO as diesel blends in the prior literature establishes the novelty of the present study.

4.5 References

[1] Kathirvel, S., Layek, A., Muthuraman, S. (2016), "Exploration of waste cooking oil methyl esters as fuel in compression ignition engines: a critical review", Eng Sci Technol Int J, v. 19, pp. 1018-1026.

[2] Xue, J. (2013), "Combustion characteristics, engine performances and emissions of waste edible oil biodiesel in diesel engine", Renew Sustain Energy Rev, v. 23, pp. 350–365.

[3] Capuano, D., Costa, M., Di Fraia, S., Massarotti, N., Vanoli, L. (2017), "Direct use of waste vegetable oil in internal combustion engines", Renew Sustain Energy Rev, v. 69, pp. 759-770.

[4] Holman, J. P. (2004), Experimental techniques for engineers, 1st ed., Tata McGraw Hill, New Delhi.

[5] Che Mat, S., Idraos, M. J., Hamid, M. F., Zainal, Z. A. (2018), "Performance and emissions of straight vegetable oils and its blends as a fuel in diesel engine: A review", Renew Sustain Energy Rev, v. 82, pp. 808-823.

[6] Martin, M. L. J., Geo, V. E., Nagalingam, B. (2016), "Effect of fuel inlet temperature on cottonseed oil-diesel mixture composition and performance in a DI diesel engine", J. Energy Institute, v. 90, pp. 563-573.

[7] Srithar, K., Balasubramanian, K. A., Pavendan, V., Kumar, B. A. (2017), "Experimental investigations on mixing of two biodiesels blended with diesel as alternative fuel for diesel engines", J King Saud Univ Eng Sci, v. 29, pp. 50-56.

[8] Yesilyurt, M. K. (2019), "The effects of the fuel injection pressure on the performance and emission characteristics of a diesel engine fueled with waste cooking oil biodiesel-diesel blends", Renew Energy, v. 132, pp. 649-666.

[9] Prabu, S. S., Asokan, M. A., Roy, R., Francis, S., Sreelekh, M. K. (2017), "Performance, Combustion and Emission Characteristics of Diesel Engine fueled with Waste Cooking Oil Bio-diesel or diesel blends with Additives", Energy, v. 122, pp. 638-648.

[10] Attia, A. M. A., Hassaneen, A. E. (2016), "Influence of diesel fuel blended with biodiesel produced from waste cooking oil on diesel engine performance", Fuel, v. 167, pp. 316-328.

[11] Agarwal, A. K., Dhar, A. (2013), "Experimental investigations of performance, emission and combustion characteristics of Karanja oil blends fuelled DICI engine", Renew. Energy, v. 52, pp. 283–291.

[12] Akshatha, M. (2018), "Hotels unaware that used cooking oil can yield biodiesel", Economic Times, https://economictimes.indiatimes.com/industry/energy/oil-gas/hotels-unaware-that-used-cookingoil-can-yield-biodiesel/articleshow/62418971.cms. Accessed 18 November 2018.

[13] Joshi, S., Hadiya, P., Shah, M., Sircar, A. (2019), "Techno-economical and experimental analysis of biodiesel production from used cooking oil", Bio Phys Economics Resource Quality, v. 4, pp. 1-6.

[14] Indian Oil Corporation Limited. 2020. Total Product List. January 2020. Accessed 21 January 2020.

Chapter 5

TO OPTIMIZE FUEL CONSUMPTION AND THERMAL EFFICIENCY OF A CI ENGINE FUELED WITH DIESEL-WASTE VEGETABLE OIL BLENDS

5.1 Introduction

As summarized in the previous chapter, WVO has the potential to be used as an alternative fraction in existing fossil fuels for powering a CI engine. The economics of

WVO blend fuels also make them valuable alternative fuel stocks. Thus, the next logical objective is to optimize the performance of an existing CI engine fueled with diesel blends containing WVO, obtained from varying thermal treatment durations and at different blend fractions.

Several engine parameters are assessed during the engine performance evaluation of blend fuels in a CI engine. But, the literature suggests that for a specific CI engine where engine parameters are invariant, the nature and properties of the fuel along with the applied torque govern the performance of the engine [1, 2]. The fuel consumption (FC) and thermal efficiency (TE) are the principal engine performance parameters affected by fuel properties and applied torque [3]. The modulation of the properties of blend fuel can be achieved by varying the blend proportion. Several strategies are adopted by researchers for optimizing the governing factors of the performance parameters of CI engines. The blend proportion is the factor that is mostly studied for the impact on engine performance parameters. A recent literature review has suggested the same [4]. The statistical approach is obligatory for the simultaneous evaluation of multiple factors at various levels. In this regard, the statistical response surface (RS) optimization is a widely adopted strategy for systematic multi-factor optimization. The optimization of WVO-diesel blend properties to engine performance was not performed previously. The pertinence of the present study and the novelty of the planned objectives are established from the state-of-the-art analysis of the relevant literature reports.

Accordingly, the present study aims to systematically evaluate the impact of WVO fraction in the diesel blend, thermal exposure time of WVO, and the applied torque (load) on engine performance parameters, namely, FC and TE, using an RS approach. Also, optimization of the factors to deliver a diesel-WVO blend with FC and TE comparable to diesel is another objective of the study. In contrary to the majority of the related work, where WVO was trans-esterified to methyl esters before blending, the present study aims to blend chemically unmodified WVO with diesel through systematic parametric optimization which can be utilized to fuel a light-duty CI engine.

5.2 Methodology

5.2.1 Preparation of WVO-diesel blends

The diesel and vegetable oil for the study was locally obtained. The waste frying oil (WFO) was collected from a single local source processing potatoes in a manual frying unit using the same vegetable oil. The oil temperature of the frying was recorded to be $150 \pm 5°C$. The variety of oil (soybean oil) was kept uniform throughout the study. The WVO for the study was made in-house by subjecting the vegetable oil to thermal treatment or different duration simulating the food processing operations, similar to frying. The temperature of thermal treatment was maintained invariant at 150 ± 5 °C, per several earlier studies [5]. As reported elsewhere, the duration of thermal treatment of oil generally varies between 24-72 hours [6]. By varying the thermal treatment duration, the different WVO stocks were prepared and stored for further characterization and blending.

5.2.2 Characterization of fuels

The density, viscosity, cetane index, and calorific value were measured using ASTM D1217, D2783, D976, and D240 standard methods, respectively. The flash and fire points were determined using the ASTM D93 method. These ASTM test methods are already discussed in section 3.2.3 in Chapter 3. The WVO of different thermal exposure duration was analyzed using Fourier transformed infrared (FT-IR) spectrophotometer (Nicolet iS10, Thermo Scientific Inc., USA). The attenuated total reflectance technique was used for generating the IR spectra of fuel samples. The peak matching technique was used to determine the changes in the functionalities of various WVO. The fatty acid profile of WVO was determined by gas chromatography (Clarus-600, Perkin Elmer, Singapore) coupled with mass spectrometry (Clarus-560, Perkin Elmer, Singapore) using the analytical technique reported in the literature [7].

5.2.3 Characterization of engine performance and emission

The same 3.7 kW CI engine, as described in section 3.2.4 in Chapter 3, was employed in this study. The engine specifications and emission uncertainties are thus listed in Table 3.3 and 3.4, respectively.

The density (g/ml) (as ρ_f) and fuel flow rate (ml/h) (as $\dot{m}_f$) were used in calculating the fuel consumption (FC) (kg/h) using the formula given in Eq. (5.1).

$$FC\ (kg/h) = [\dot{m}_f(ml/h) * \rho_f\ (g/ml)]/\ 1000 \tag{5.1}$$

The thermal efficiency (TE) was calculated using fuel consumption (FC), the power (kW) produced at a specific engine torque (shown as P), and higher heating value (kJ/kg) of the fuel (shown as HHV), using the formula given by [8], represented Eq. (5.2).

$$TE\ (\%) = [P\ (kW)*3600] / [(HHV\ (kJ/kg)*FC\ (kg/h))*100] \tag{5.2}$$

The mechanical efficiency (ME) of an engine is the ratio of the output power transmitted to the engine shaft and the input power at the applied engine torque.

5.2.4 Experimental design and model development using RS approach

5.2.4.1 Design factors and levels

The effect of WVO fraction in the blend, thermal treatment time of WVO, and engine torque on FC and TE of CI engine is systematically evaluated in an RS approach using experimental design. The three design factors were: WVO fractions in the blend, thermal treatment time of WVO, and engine torque. The levels of the three factors were determined based on accessible resources and results from pilot experiments. The thermal treatment time of WVO varied between 24h and 72h. The maximum thermal exposure of WVO was restricted to 72h because the thermal exposure of vegetable oil in food processing operation rarely extends beyond continuous three days [6]. Applied engine torque due to load on the engine varied between 4.91 Nm to 14.72 Nm. The engine torque beyond 14.72 Nm could not be evaluated for the test engine as it resulted in the abrupt stopping of the engine. The fraction of WVO in the diesel blend varied between 0.20 and 0.50. As the test engine consisted of a conventional fuel injector, a WVO fraction greater than 0.50 in diesel was not considered in the study. For the optimization of three or more factors, Box-Behnken Design (BBD) is a widely preferred experimental design, owing to its wider orthogonality, better predictability, and fewer experimental requirements [9]. BBD is discussed in section 3.2.4.2 in Chapter 3. The FC for the CI engine was modeled in the RS approach using BBD experimental array. The details of the factors-levels of the experimental array are presented in Table 5.1.

Table 5.1 Factors and levels of the experimental array.

Factor levels	Thermal treatment time (h)	WVO fraction in the blend	Engine torque/ load (Nm)
1	24	0.20	4.91
2	48	0.35	9.81
3	72	0.50	14.72

5.2.4.2 Development of RS model

A non-linear quadratic model for the response (FC) was evaluated. A quadratic correlation has been used because it contains not only the single factor effects (linear terms) but also the combined effects of two factors at a time and the squares of each factor. Thus, a quadratic correlation is more robust than a linear correlation. Additionally, the selection of a quadratic correlation for regression-based models is mostly preferred by researchers throughout the world [3]. The quadratic model (Eq. (5.3)) expressed the FC as a function of the WVO fraction in the blend, thermal treatment time of WVO, and engine torque (factors). In Eq. (5.3), $A_0 - A_9$ are the regression coefficients for the respective model terms. ε is the "error" in the model.

$$FC \text{ (kg/h)} = A_0 + A_1(\text{WVO fraction}) + A_2(\text{thermal treatment time}) + A_3(\text{engine torque}) + \tag{5.3}$$
$$A_4(\text{WVO fraction})^2 + A_5(\text{thermal treatment time})^2 + A_6(\text{engine torque})^2 +$$
$$A_7(\text{WVO fraction in blend})(\text{thermal treatment time}) + A_8(\text{WVO fraction in}$$
$$\text{blend) (engine torque)} + A_9(\text{engine torque})(\text{thermal treatment time}) + \varepsilon$$

Table 5.2 presents the response to the experimental array. The FC of diesel blends for each design point (experiment) was measured. Fifteen experiments, performed in triplicates, were marked as blocks. The replication facilitated in determining the experimental error.

Table 5.2 The response at different factor level settings of the experimental design array.

Exp. Order	Experimental factors			Response (FC (kg/h))		
	Thermal treatmen t time (h)	WVO fraction in blend	Engine torque/ load (Nm)	Rep 1 (kg/h)	Rep 2 (kg/h)	Rep 3 (kg/h)
1	50	24	9.81	0.647	0.661	0.634
2	80	24	9.81	0.603	0.601	0.606
3	50	72	9.81	0.662	0.671	0.662
4	80	72	9.81	0.508	0.509	0.507
5	50	48	4.91	0.506	0.505	0.504
6	80	48	4.91	0.502	0.502	0.504
7	50	48	14.72	0.509	0.509	0.509
8	80	48	14.72	0.501	0.496	0.502
9	65	24	4.91	0.621	0.629	0.620
10	65	72	4.91	0.608	0.611	0.606
11	65	24	14.72	0.638	0.647	0.631
12	65	72	14.72	0.643	0.660	0.639
13	65	48	9.81	0.652	0.690	0.631
14	65	48	9.81	0.681	0.696	0.672
15	65	48	9.81	0.686	0.694	0.673

The order and coefficients of the RS model were concluded from the analysis of the variance of the experimental response. The coefficients of the RS model (Eq. (5.3)) for the FC for the blended fuel (response) as a function of the WVO thermal treatment time, the fraction of WVO in the blend, and torque on the engine (the process parameters), were established by RS approach from regression analysis using MINITAB® version 16 statistical software (Minitab Inc., State College, PA, USA). The comparison of the t-values with the critical value established the statistical significance of the coefficients [10]. The final RS model was defined through the backward elimination method by deleting the statistically insignificant terms. A similar methodology was applied to develop an RS model for TE. Additional experiments were conducted for the three variables under consideration to validate the RS model. The most popular criterion of optimization of an RS model is the D-optimality criterion [10]. The maximum response within the factor space for minimizing FC and maximizing TE was located using the D-optimality-based numerical optimization function. The desirability test is one of the most widely-used methods to optimize responses based on a few fixed factors. For a response to be the desired one, the factors and their levels need to be finely tuned. Before tuning the factors and their associated levels, it has to be made sure whether the response should be maximized, minimized, or kept unchanged. Thus, to obtain the desired response the factors within the extremities of their levels were given weightage from 0.1 to 1.0. Where 0.1 is the lowest weightage and 1.0 being the highest one. A similar weightage is given to the response, which is known as the desirability or D-optimality value. The desired response was to minimize FC and maximize TE.

5.2.5 Real-life evaluation

The suitability of the optimized factors established in the study was verified using waste frying oil (WFO) collected from a single local source where similar vegetable oil (soybean oil) is used for frying purposes. The total frying cycles of the oil correspond to 47±3 h of thermal exposure. The frying temperature was 150±5 °C, similar to those reported previously [6]. The collected WFO was filtered and mixed with diesel at the optimized blend ratio. The diesel-WFO (D-WFO) blend was characterized for the fuel properties following ASTM standard test methods. The engine performance and emission studies were also conducted with the D-WFO blend and compared with the model-optimized D-WVO blend.

5.3 Experimental findings

5.3.1 Characterization and analysis of fuels

The complete fatty acid profile of the three WVO stocks is presented in Table 5.3. All WVO-BD stocks have a high compositional presence of polyunsaturated fatty acids (PUFA) (linoleic acid (C18:2) and linolenic acid (C18:3)), followed by monounsaturated fatty acids (MUFA) (oleic acid (C18:1)). Palmitic acid (C16:0) and stearic acid were the abundant saturated fatty acids (SFA) in WVO. The fatty acid profiles of different WVO-BDs are in agreement with those reported in the literature [12]. The free fatty acid content of the WVO stocks varied between 1.01 - 1.06 % (w/w). The duration of thermal treatment has resulted in a general decrease in the unsaturation in the oil, evident from the rise in the total SFA. The loss in unsaturation is linked with the formation of polymerized and conjugated products in the oil, which in turn impacts the physical properties of the oil [6, 11].

Table 5.3 Fatty acid composition of WVO of varying thermally exposure durations.

Fatty acid	MW (g/mol)	Structure	Formula	WVO_24h (% w/w)	WVO_48h (% w/w)	WVO_72h (% w/w)
Myristic acid	228	C14:0	$C_{14}H_{28}O_2$	0.14	0.09	0.02
Palmitic acid	256	C16:0	$C_{16}H_{32}O_2$	12.39	13.21	13.51
Stearic acid	284	C18:0	$C_{18}H_{36}O_2$	9.02	9.56	9.81
Oleic acid	282	C18:1	$C_{18}H_{34}O_2$	36.21	35.84	34.98
Linoleic acid	280	C18:2	$C_{18}H_{32}O_2$	41.18	40.27	40.12
Linolenic acid	278	C18:3	$C_{18}H_{30}O_2$	0.27	0.23	0.21
Total SFA				21.55	24.15	23.40
Total MUFA				36.21	34.23	34.98
Total PUFA				41.43	41.19	40.34
Free fatty acid				1.01	1.04	1.06

Note: MW – molecular weight
SFA - saturated fatty acid (CX:0)
MUFA – monounsaturated fatty acid (CX:1)
PUFA – polyunsaturated fatty acid C(X:2), and C(X:3).
where, X denotes carbon number; 0, 1, 2, and 3 denotes the respective number of double bonds.

The characterization results of the diesel, WVOs subjected to different thermal treatment duration, and WVO-diesel blends are reported in Table 5.4. The density and viscosity of WVO increased with increasing thermal treatment time, as shown in Table 5.4. The increment in density and viscosity of WVO with prolonging thermal treatment duration may be contributed to amplified polymerization products in WVO. Increased thermal treatment durations also increased the flash and fire points of WVO. The WVO subjected to prolonged thermal treatment time becomes unusable as an alternative fuel for the CI engine. Earlier studies reported that increased polymerization products in WVO subjected to long thermal exposure reduce the transesterification potential of WVO for utilization as a diesel fuel alternative [11]. FT-IR spectra (Fig. 5.1) revealed that WVO tends to form secondary products of polymerization on thermal exposure of

48h and beyond. Distinct changes in the functionalities were observed between the wavenumber 1450 to 1700 cm⁻¹. The FT-IR spectra confirmed a decrease in C=O functionalities with increased thermal exposure duration. The calorific values for the different WVOs were also significantly lower than the diesel fuel. An increase in calorific value of WVO on prolonging thermal exposure corroborates the findings reported in Fig. 5.1. The WVO was well-miscible with diesel. The density, kinematic viscosity, flash point, fire point, cetane index, and calorific value of different diesel-WVO blends were measured.

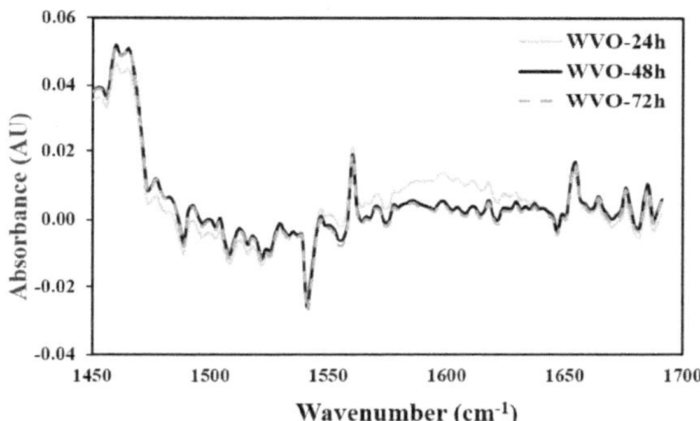

Fig. 5.1 FT-IR spectra of different WVO (after a different degree of thermal exposure).

Table 5.4 The results of characterization of various fuels evaluated.

Fuel samples	Density at 15 °C (g/cc)	Viscosity at 20 °C (cSt)	Flash point (°C)	Fire point (°C)	Cetane Index	Calorific value (kJ/kg)
Diesel fuel	0.823	3.38	48	58	52.9	46,656
WVO_24h	0.875	57.44	178	> 200	46.1	39,660
WVO_48h	0.881	67.66	189	> 200	46.2	39,990
WVO_72h	0.887	70.01	> 200	> 200	46.3	40,126
D-WVO_48h_0.20[a]	0.841	6.01	48	68	51.3	45,324
D-WVO_48h_0.35[a]	0.851	9.36	49	70	50.2	44,374
D-WVO_48h_0.50[a]	0.861	15.24	55	82	49.7	43,802
Test Method	ASTM D1217	ASTM D2983	ASTM D93		ASTM D976	ASTM D240

Note: [a] - WVO-diesel blends shown only for WVO_48h oils.

All WVO blends showed comparable densities to the range $0.82 - 0.86$ g/cm^3, as specified for diesel fuel by IS 1460 standards. Increasing WVO fraction in the blend increased the kinematic viscosity of the blended fuel. The kinematic viscosity for the blends varied between 6.193 and 15.237 cSt and exceeded the standard range for diesel. The blended fuels have a flashpoint of 48.5 ± 6.5 °C and fire point of 71 ± 11 °C, respectively. The higher flash point of WVO-diesel blends compared to diesel results in lower fuel volatility. Lower fuel volatility influences the evaporation characteristics of the fuel but makes the blends safer than diesel for handling and storage. The cetane indices of all the blends were comparable to that of diesel. However, the blend cetane index decreased from 51.3 to 49.7 with an increase in the WVO fraction. A minimum variation in ignition delay may be expected due to a change in fuel for the CI engine during the study. The calorific value of different WVO-diesel blends varied between 43,802 to 45,324 kJ/kg. The calorific value of the blends increased with an increase in thermal exposure of WVO and decreased with the richness of the WVO in the blend. The calorific value of all blended fuel was close to that for diesel fuel. The comparable calorific value among D-WVO blends suggested that variations in the power output among the blended fuels for the same amount of fuel injected in the cylinder under particular engine loads may not be significant. The D-WVO blends proposed in this study holds good potential as an alternative fuel in CI engines.

5.3.2 Analysis of factor effects and development of RS model

The response variable (FC) was determined at each design point of the experimental array. Fig. 5.2 illustrates the matrix plots for the response (FC) in a 3 factor-3 level experimental design array. Relatively low FC was recorded for increased WVO fraction (0.50 WVO in the blend). Density, viscosity, and calorific value are the fuel properties that define the FC of a CI engine. The density of fuel controls fuel atomization, droplet formation, and penetration of the fuel jet inside the cylinder. While proper viscosity of fuel prevents fuel loss during the injection process. Proper density and viscosity of fuel are requisite in maintaining optimum FC in a CI engine. Similar explanations are present in the literature [13]. The calorific value is the defining factor for the combustion of fuel inside the cylinder. A minor change in calorific value may not cause a drastic change in the combustion outcome, therefore may not result in a major rise in FC. Although the higher densities of the various D-WVO_0.50 blends caused higher mass injection for the same volume of fuel injected, due to higher viscosities, there may have been a reduction in fuel loss, proper combustion of the injected fuel mass, and faster evolution of pressure inside the cylinder. Thus, a lower FC for 0.50 WVO fraction in the blend may be due to an optimum balance between droplet atomization and penetration of fuel jet under increased fuel density and viscosity.

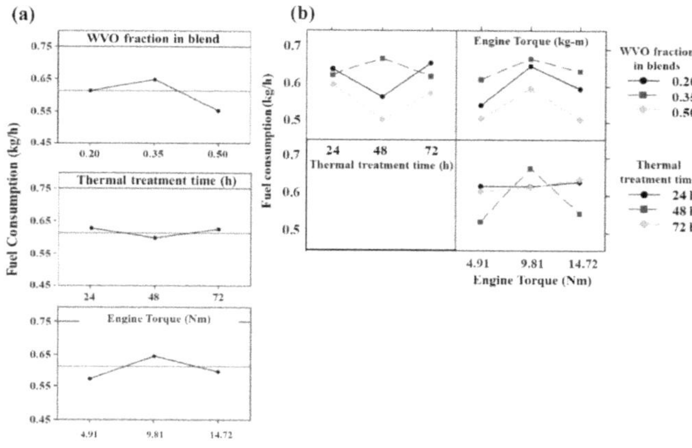

Fig. 5.2 Matrix of plots of experimental factors for fuel consumption in three factors - three levels BBD. **(a)** Main effects plot – mean response of a factor [factors other than the main factor were held at a mean level] **(b)** Two-factor interaction plots – interaction among factors at the associated levels.

Fig. 5.3 illustrates the surface plots of the response (FC (kg/h)) for the experimental factors. Fig. 5.3(a) shows that a blend of 48h thermally treated WVO in fractions of 0.50 and 0.20 resulted in a low FC. A higher calorific value of the WVO_0.20 blend is a probable reason for lower FC. An increase calorific value enhances fuel combustion inside the cylinder which causes lesser fuel to be consumed for generating the desired power. Whereas, elevated density and viscosity of WVO may have resulted in a reduced FC of the diesel blend with a higher WVO fraction. Also, higher fractions of oxygenated fuel (with carboxylic functionalities) are known to improve the combustion properties of the fuel blend [2]. Thus, higher WVO content in the blend may have resulted in better combustion and hence minimized FC. However, the magnitude of change in the FC due to thermal treatment of oil was much lower compared to the change in FC caused by varying blend proportions or engine torque.

The surface plot for WVO fraction in the blend versus engine torque (Fig. 5.3(b)) shows that low FC resulted in blends with 0.50 WVO fraction for all engine torque settings. A fine balance between fuel density and calorific value, as discussed earlier, maybe a probable reason for the trend. An increase in fuel density and fuel viscosity with a marginal change in heating value for higher WVO fraction in the blend may have favored optimum droplet atomization and jet penetration, thereby minimized the FC. With an increase in engine torque, FC was seen to reduce. It is because at high engine torque/ load due to high in-cylinder temperature, combustion is facilitated and comparatively lower fuel mass is required to generate the desired work.

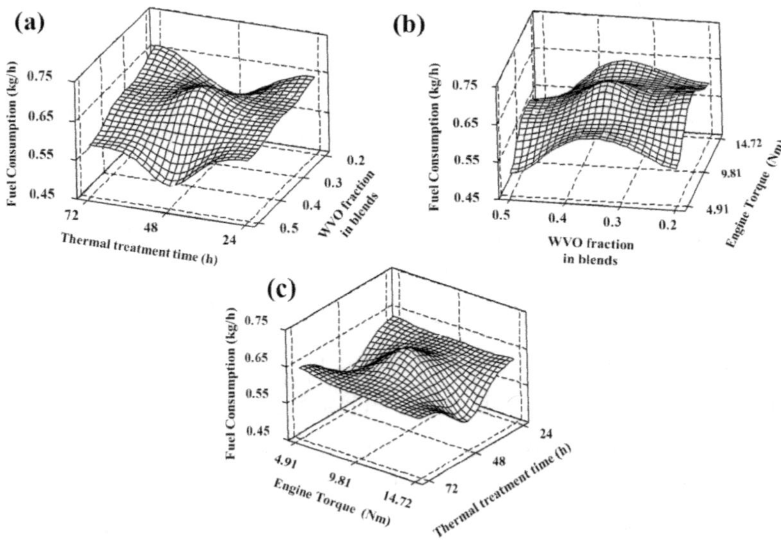

Fig. 5.3 Surface plots of the response (FC (kg/h)) for the experimental factors **(a)** WVO fraction in the blend and WVO thermal treatment time (h), **(b)** WVO fraction in the blend and engine torque (Nm), and **(c)** Engine torque (Nm) and WVO thermal treatment time (h).

The response surface in Fig. 5.3(c) shows that for WVO thermal treatment time of 48h, at both low and high engine torques, FC was lower as compared to other combinations. For 24h and 72h treated WVO, changes in engine torque did not cause any drastic increase in FC. This strange trend suggests that WVO, which is thermally treated for 48h is the most suitable WVO to be blended with diesel for fueling a CI engine. Also, the FC was lower for a higher engine torque setting (14.72 Nm) as compared to the lower torque settings.

Table 5.5 The results of the analysis of variance of the experimental response.

Terms		DF	Sum of Squares	Mean Sum of Squares	F	p (F>F$_{0.05}$)
Blocks		2	0.001558	0.000779	6.08	0.060
Regression		9	0.130549	0.014505	113.21	0.000*
Linear		3	0.025483	0.025748	200.96	0.000*
	WVO fraction in blend	1	0.022707	0.042448	331.29	0.000*
	Thermal treatment time	1	0.000043	0.001908	14.89	0.001*
	Engine torque	1	0.002733	0.035271	275.28	0.000*
Square		3	0.101939	0.033980	265.20	0.000*
	(WVO fraction in blend)^2	1	0.049457	0.053820	420.05	0.000*
	(Thermal treatment time)^2	1	0.006382	0.003986	31.11	0.000*
	(Engine torque)^2	1	0.046100	0.046100	359.80	0.000*
Interaction		3	0.003127	0.001042	8.14	0.000*
	WVO fraction in blend x Thermal treatment time	1	0.001186	0.001186	9.26	0.005*
	WVO fraction in blend x Engine torque	1	0.001521	0.001521	11.87	0.002*
	Thermal treatment time x Engine torque	1	0.000420	0.000420	3.28	0.079
Residual Error		33	0.004228	0.000128		
	Lack-of-Fit	27	0.002383	0.000088	0.29	0.989
	Pure Error	6	0.001845	0.000308		
Total		44	0.136335			

Note: DF = degrees of freedom; SS = sequential sum of square; MS = adjusted mean of square; (*) = values are statistically significant at 5% level of significance; p = a statement describing F.

The RS model Eq. (5.3) was refined to exclude the terms that are insignificant (statistically). The significance of the terms was arrived at from the analysis of variance results tabulated in Table 5.5. The net error of the experiment for the design array was assessed from the "lack-of-fit" and the error contribution was confirmed as insignificant (p-value (0.989) > 0.05) [10]. Eq. 5.4 represents the refined RS model comprising of significant terms. Also, Table 5.6 denotes the coefficients of the various terms in the quadratic RS model for FC. Those coefficients which have a p-value > 0.05 were considered insignificant and are removed from the refined RS model.

$$FC\ (kg/h) = 0.091140 + 2.24619*(WVO\ fraction) - 0.065860*(thermal\ treatment\ time) \quad + \quad (5.4)$$
$$0.0577278*(engine\ torque) - 3.097920*(WVO\ fraction)^2 + 0.018970*(thermal$$
$$treatment\ time)^2 - 0.0026814*(engine\ torque)^2 - 0.066280*(WVO$$
$$fraction)*(thermal\ treatment\ time) - 0.0153007*(WVO\ fraction)*(engine\ torque)$$

Table 5.6 The values of the coefficients of the RS model.

Terms	Coeff.	Reg. coeff.	T	$p(T>T_{0.05})$
Constant	0.091140	0.036735	2.481	0.018*
Blocks	0.000320	0.002386	0.135	0.894
WVO fraction in blend	2.246190	0.123407	18.201	0.000*
Thermal treatment time	(-) 0.065860	0.017066	(-) 3.859	0.001*
Engine torque	0.0577278	0.003473	16.592	0.000*
(WVO fraction in blend)^2	(-) 3.097980	0.151157	(-) 20.495	0.000*
(Thermal treatment time)^2	0.018970	0.003401	5.578	0.000*
(Engine torque)^2	(-) 0.0026814	0.000144	(-) 18.968	0.000*
WVO fraction in blend x Thermal treatment time	(-) 0.066280	0.021784	(-) 3.043	0.005*
WVO fraction in blend x Engine torque	(-) 0.0153007	0.004442	(-) 3.445	0.002*
Thermal treatment time x Engine torque	0.0012059	0.000662	1.811	0.079

Note: p = a statement which describes T, Reg. coeff. = regression coefficient
(*) = values are statistically significant at the 5% level of significance.
Unit: Thermal treatment time in hours (h) and Engine torque in Newton-meter (Nm).

5.3.3 Validation of the RS model

FC computed using the model equation (Eq. (5.4)) was plotted against the experimentally obtained FC. Fig. 5.4(a) revealed a reasonable correlation ($R^2 = 0.958$) between the experimentally obtained FC and model-predicted values of FC. A paired t-test was performed with model-predicted FC and experimental FC to assess the significance of the difference between the data set. The paired t-test confirmed that the difference between the experimental FC and the model-predicted FC was not statistically significant ($t_{calculated}$ (0.04) < $t_{tabulated}$ (2.92)).

The accuracy of the RS model at factor settings other than the design points was evaluated using a validation study. Fig. 5.4(b-d) represented the validation study. The experimental FC values were close to the model-predicted values for WVO fractions of 0.20 and 0.35 (Fig. 5.4(b)). At a higher WVO fraction, the experimental FC was slightly higher than that predicted by the model; however, the difference was confirmed to be statistically insignificant. The experimentally obtained FC values for diesel blends with WVO exposed to varied thermal treatment durations were plotted against model-predicted FC values. The model prediction closely matched the experimental values, presented in Fig. 5.4(c). The experimental value of FC was comparable with the model-predicted value for all settings of engine torque (Fig. 5.4(d)) except for the higher engine torque settings. The experimental values at different torque settings followed an identical trend as predicted by the model. Thus, the model for FC was confirmed to be valid for different factor level settings within the factor space considered in this study.

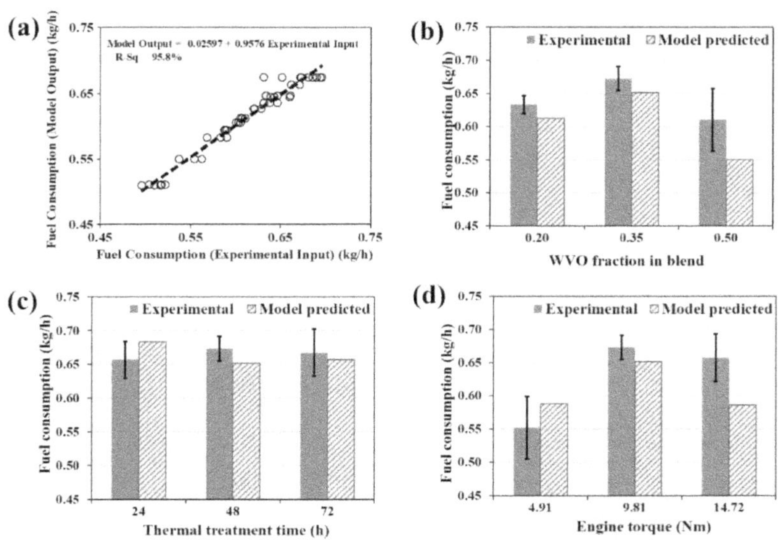

Fig. 5.4 Validation study of the RS model - **(a)** Correlation plot of experimental *vs.* Model-predicted FC, **(b)** validation for different WVO fraction, **(c)** validation for various WVO thermal treatment times, and **(d)** validation for change in engine torque.

5.3.4 Extension of RS model for TE

The thermal efficiency (TE) for a CI engine depends on the fuel consumption rate and the engine output power at a specific torque. Additional experiments were conducted with the response surface model for FC (Eq. (5.4)) to define the model for predicting the TE of the CI engine. Eq. (5.5) presents the model equation for TE. In Eq. (5.5) WVO fraction is unitless, thermal treatment time is in hours (h), and engine torque is in Newton-meter (Nm). The model (Eq. (5.5)) was used to evaluate the effect of the experimental factors at various factor levels on the TE of the CI engine. Fig. 5.5 represents the surface plots for TE at different factors-levels.

$$
\begin{aligned}
\text{TE} = \; & 0.354111 - (0.687334*\text{WVO fraction}) + (0.020153*\text{thermal treatment time}) \\
& - (0.017665*\text{engine torque}) + (0.947964*(\text{WVO fraction})^2) \\
& - (0.005805*(\text{thermal treatment time})^2) + (0.000821*(\text{engine torque})^2) \\
& + (0.020282*(\text{WVO fraction})*(\text{thermal treatment time})) \\
& + (0.004682*(\text{WVO fraction})*(\text{engine torque}))
\end{aligned} \tag{5.5}
$$

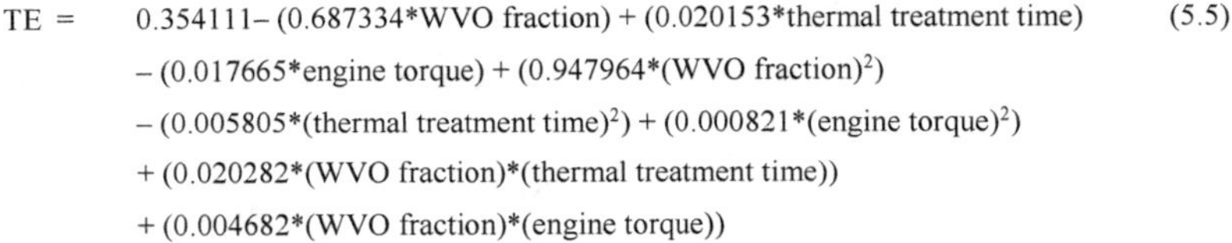
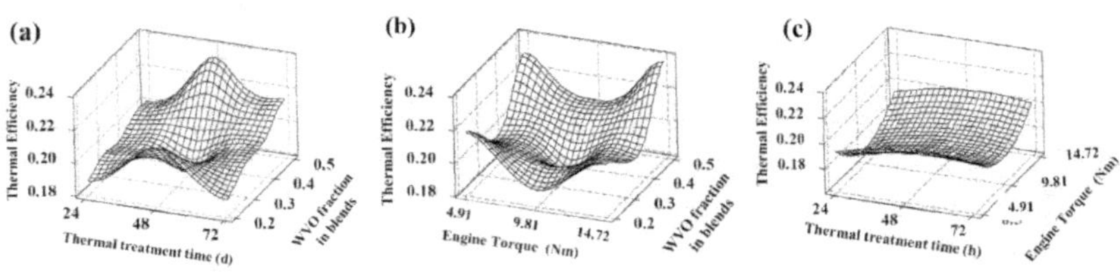

Fig. 5.5 Three-dimensional surface plots of the response variable (thermal efficiency) for the different experimental factors (two-factor-at-a-time).

Fig. 5.6 presents the validation study for TE. The experimental TE values closely matched the model-predicted TE for all WVO fractions in the blend. Similar accuracy of the model prediction of TE was noted for thermal treatment time (Fig. 5.6(b)). The experimental TE values were comparable with the model-predicted TE values for different torque settings (Fig. 5.6(c)). Except for engine torque of 9.81 Nm (middle setting), high variability in experimental TE was recorded for other engine torque settings (4.91 and 9.81 Nm). High fluctuations in FC at high and low engine torque may be attributed to the recorded variability in TE. The model for TE was confirmed to be valid for different factors-levels within the factor space considered in this study.

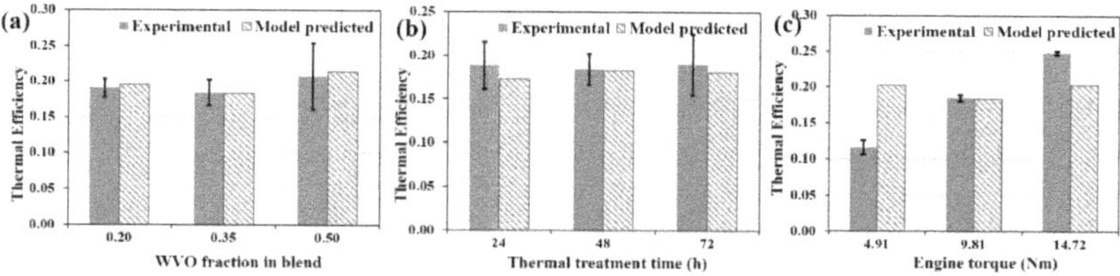

Fig. 5.6 Validation study of the RS model extension for TE **(a)** validation for different WVO fractions, **(b)** validation for various WVO thermal treatment times, and **(c)** validation for changes in engine torque.

5.3.4 Optimization of FC and TE

A numerical optimization analysis using the D-optimality criterion was performed to minimize FC and maximize TE. The factor settings were tuned between the worst case (zero) and the ideal case (one) of the optimality index for delivering the optimum response [10]. Table 5.7 tabulates the effect of varying factors-levels on the D-optimality criterion for FC and TE.

Table 5.7 Simulation results for varying the factors to improve the fuel consumption, thermal efficiency of the engine, and D-optimality criterion.

WVO fraction in the blend	Thermal treatment time (h)	Engine torque / load(Nm)	Model FC (kg/h)	Model TE (%)	D-optimality
0.50	45	14.72	0.5011	23.58	0.937
0.50	46	14.72	0.5010	23.62	0.988
0.50	47	14.72	0.5097	23.66	1.000
0.50	48	14.72	0.5094	23.70	1.000
0.50	**54**	**14.72**	**0.5092**	**23.84**	**1.000**
0.50	48	14.62	0.5011	23.66	0.931
0.49	48	14.72	0.5011	23.66	0.939
0.50	48	4.91	0.5099	22.94	1.000
0.50	54	4.91	0.5075	23.08	1.000
0.50	60	4.91	0.5067	23.15	1.000
0.50	66	4.91	0.5083	23.15	1.000

The blend containing 0.50 fraction as 47h thermally treated WVO at an engine torque of 14.72 Nm recorded as the optimum factor-levels for the maximum response (ideal case, D-optimality = 1.0). Lower WVO fraction in diesel blend lowered the D-optimality criterion. An increase in thermal treatment time (up to 54h) marginally increases the D-optimality criterion. The D-optimality of 1.0 was computed both at high (14.72 Nm) and low (4.91 Nm) engine torque. The overlaid contour plots of FC and TE were evaluated to confirm the region of optimum response. Fig. 5.7(a) showed that high engine torque had a more desirable region for an optimum response of lower FC and higher TE. Fig. 5.7(b) identified that high WVO fraction thermally treated for 48h and beyond defines the factor level setting of the desired region for lower FC and higher TE. The minimum FC was 0.5010 kg/h and the maximum TE was 23.84%. The experimentally obtained FC at factor setting nearest to the optimum recorded a value of 0.5094 kg/h and that for TE was 23.70%.

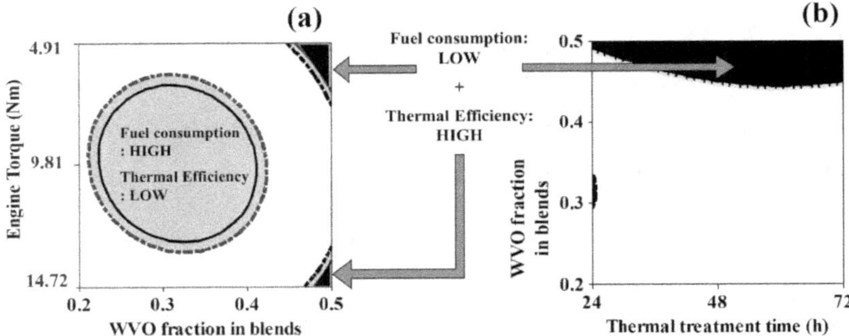

Fig. 5.7 Overlaid contour plots showing the desirability region **(a)** for various WVO fractions in the blend and engine torque (Nm), and **(b)** for various WVO fractions in the blend and thermal treatment time (h).

Table 5.8 Comparison of the FC and TE of various fuel blends applied in a CI engine at or around 14.72 Nm engine torque/load (or 2.5 kW brake power).

Fuel blends	FC (kg/h)	TE (%)	References
Straight vegetable oils and diesel blends			
100% Soybean oil	0.896	31.0	Nettles-Anderson 2014 [14]
100% Sunflower oil	0.868	33.0	Nettles-Anderson 2014 [14]
100% Canola oil	0.882	31.3	Nettles-Anderson 2014 [14]
100% Papaya seed oil	1.014	25.0	Ashokan et al. 2018 [15]
100% Watermelon seed oil	0.858	24.0	Ashokan et al. 2018 [15]
50% Karanja oil + 50% diesel	1.554	23.6	Agarwal and Dhar 2013 [13]
40% Cashewnut shell oil + 40% diesel	N/A	25.3	Bupeshraja et al. 2019 [16]
60% Cotton seed oil + 40% diesel	N/A	27.0	Martin et al. 2016 [17]
40% Mahua oil + 60% diesel	0.688	27.2	Sonar et al. 2015 [18]
40% Poon oil + 60% diesel	0.990	18.8	Devan and Mahalakshmi 2009 [19]
Vegetable oil biodiesels and diesel blends			
20% Castor oil biodiesel + 80% diesel	0.773	23.8	Das et al. 2017 [20]
25% Karanja oil biodiesel + 75% diesel	1.620	24.5	Shivaramakrishnan 2017 [21]
30% Tamanu oil biodiesel + 70% diesel	0.801	27.5	Miraculas 2016 [22]
60% Rubber seed oil biodiesel + 40% diesel	0.844	29.1	Murugapoopathi 2018 [23]
30% Neem oil biodiesel + 70% diesel	0.798	22.5	Datla et al. 2019 [24]
Diesel with alcohols as additives			
10% Methanol + 90% diesel	1.100	18.5	Emiroglu and Sen 2018 [25]
10% Ethanol + 90% diesel	0.751	21.8	Emiroglu and Sen 2018 [25]
30% Butanol + 70% diesel	0.656	25.8	Saxena and Maurya 2017 [26]
WVO biodiesels and diesel blends			
30% WVO Biodiesel + 70% diesel	1.003	21.5	Yesilyurt 2019 [27]
20% WVO Biodiesel + 80% diesel	0.705	28.7	Balakumar et al. 2019 [28]
40% WVO Biodiesel + 60% diesel	1.001	24.5	Prabu et al. 2017 [7]
30% WVO Biodiesel + 50% diesel	1.107	26.5	Attia and Hassaneen 2016 [29]
20% WVO Biodiesel + 80% diesel	0.688	30.7	Al-Dawody et al. 2019 [30]
WVO-diesel blends			
50% WVO + 50% diesel	0.509	23.7	Present
50% WFO [#] + 50% diesel	0.529	24.3	Study

Note: N/A - Not available, WCO – waste cooking oil [#] - real-life waste frying oil (WFO)

Table 5.8 demonstrates a comparison of the experimentally obtained optimum FC and TE with that reported in the literature. It is noteworthy to state that no study is reported to date where chemically unmodified waste vegetable oil blended with diesel was optimized for predicting engine performance. This fact established the novelty of the present study. TE reported in the present study was comparable with literature values. However, the FC of the present study was significantly lower than similar diesel fuel blends reported in the earlier studies. Hence, the optimized blend of thermally exposed WVO with diesel reported in the present study recorded a fuel blend that is more economical than the alternatives reported in the literature. In addition to providing an economical alternative fuel for CI engines, the present study also presents a sustainable utilization route for an otherwise problematic liquid waste, WVO.

5.3.5 Real-life evaluation of engine performance and emission

A comparison of the physical properties for the D-WFO blend and the optimized D-WVO blend and diesel is presented in Table 5.8. The densities of the WFO and WVO blends nearly matched. The densities for blends were comparable with the standard value of diesel. A marginally higher density of WFO may be due to the dissolved food ingredients. The cetane index and calorific value of D-WFO and D-WVO blend were much comparable. The kinematic viscosity of WFO and WVO blend with diesel were significantly higher than diesel. The properties of simulated WVO as a blend with diesel were equivalent to real-life WFO.

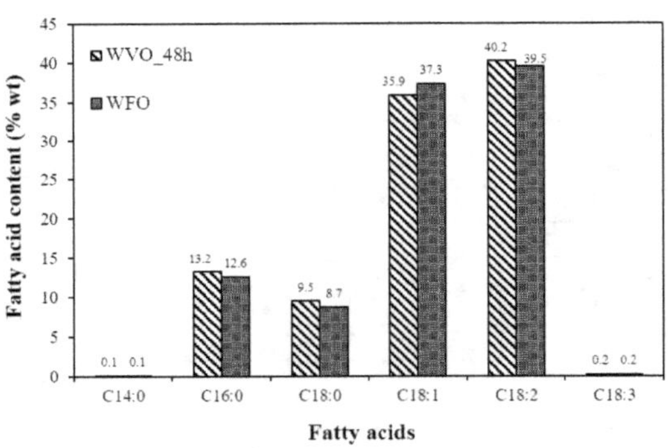

Fig. 5.8 Fatty acid profile of WVO_48h and WFO as determined from GC analysis.

The GC analysis revealed that the predominant fatty acids present in WVO stock were: palmitic acid (C16:0), stearic acid (C18:0), oleic acid (C18:1), and linoleic acid (C18:2), whose content varied from 13-13.3%, 9.1-9.6%, 35.6-37.3%, and 40-40.3%, respectively. The unsaturated fatty acids content in the WVO stocks varied between 75-76%. The WFO showed a very similar fatty acid profile with WVO having 12.6% palmitic, 8.7% stearic acid, 37.3% oleic acid, and 39.5% linoleic acid with an unsaturated fatty acid content of 77%. The fatty acid profiles of the oil stocks were in concurrence with the literature [12]. The comparison of the fatty acid profile further established the simulated WVO stock used in the diesel blend was equivalent to the real-life WFO obtained from commercial establishments. Fig. 5.8 shows the fatty acid content of WVO_48h and WFO.

Table 5.9 Comparison of properties, engine performance, and emission characteristics between optimized WVO blend and WFO blend against diesel.

Fuel properties					
Fuel samples	**Density at 15°C (g/cc)**	**Viscosity at 20°C (cSt)**	**Flash point (°C)**	**Cetane index**	**HHV (kJ/kg)**
Diesel	0.823 ±0.008	3.38 ±0.05	48 ±2	52.9 ±0.2	46,656 ±74
D-WVO_0.50	0.861 ±0.009	15.24 ±0.07	55 ±4	50.2 ±0.3	43,802 ±88
D-WFO_0.50	0.865 ±0.010	12.87 ±0.09	67 ±4	49.8 ±0.3	43,426 ±85
Engine performance characteristics at 60% (or 2.5 kW BP) engine load					
Fuel samples	**FC (kg/h)**		**TE (%)**		**ME (%)**
Diesel	0.524 ±0.031		24.06 ±0.01		80.64 ±0.45
D-WVO_0.50	0.509 ±0.012		23.57 ±0.49		82.07 ±0.12
D-WFO_0.50	0.529 ±0.008		24.28 ±0.54		81.15 ±0.25
Engine emission characteristics at 60% (or 2.5 kW BP) engine load					
Fuel samples	**CO (%)**	**CO_2 (%)**		**HC (ppm)**	**NO_x (ppm)**
Diesel	0.035 ±0.007	4.65 ±0.495		24 ±2.828	687 ±21.213
D-WVO_0.50	0.030 ±0.006	3.90 ±0.141		26 ±3.536	584 ±12.728
D-WFO_0.50	0.035 ±0.006	4.25 ±0.354		36 ±5.657	573 ±23.335

The performance of D-WVO and D-WFO blends for the optimized conditions were evaluated at 60% engine load condition and compared with diesel fuel (Table 5.9). FC of the blended fuels was very much comparable to that of the diesel fuel. Computed t-values for FC among diesel, D-WVO blend and D-WFO blend were $t_{D-D\_WVO} = 0.78$ ($p = 0.52; > 0.05$), $t_{D-D\_WFO} = 0.27$ ($p = 0.81; > 0.05$) and $t_{D\_WVO-D\_WVO} = 2.40$ ($p = 0.10; > 0.05$). The FC values of diesel against various blends were statistically insignificant at a 95% confidence level ($p > 0.05$). The computed t-values for TE among diesel, D-WVO blend and D-WFO blend were $t_{D-D\_WVO} = 1.73$ ($p = 0.22; > 0.05$), $t_{D-D\_WFO} = 0.71$ ($p = 0.55; > 0.05$) and $t_{D\_WVO-D\_WVO} = 1.69$ ($p = 0.19; > 0.05$). The ME for the blended fuels were comparable between D-WVO and D-WFO blends as well as with that for diesel fuel. The results confirmed that the FC validated from the study for the D-WVO blend was in no way different from the D-WFO blend. The result also validated that the proposed blends have the potential to replace neat diesel as an alternate fuel for CI engines without hampering the engine performance.

The emission results of D-WVO and D-WFO blends for the optimized conditions were compared with those of diesel fuel at 60% engine load (or 2.5 kW BP) condition and tabulated in Table 5.9. The results showed that the emission of the D-WVO blend was comparable with that of the D-WFO blend except for unburnt hydrocarbon (HC). Higher HC emission was noted for the WFO-diesel blend. The HC emission for the D-WFO blend may be attributed to the presence of dissolved food ingredients in WFO. Similar NO_x emission was recorded for D-WVO and D-WFO blends. However, CO and NO_x emissions for the D-WVO blend for the optimized conditions were respectively 14.3% and 17.6% lower than that for conventional diesel fuel. The proposed D-WVO blend for the optimized conditions holds an edge over the diesel fuel in terms of performance and emission.

5.4 Summary

Optimization of the characteristics of the diesel-WVO blend was performed to minimize the FC and maximize the TE in a CI engine. A response surface (RS) based statistical approach was adopted for the optimization. A predictive model of FC was developed using an RS approach. The RS model for FC was refined to predict the TE of the CI engine within the defined factor space. The model predictions were well-correlated at all the factors-levels. Based on the D-optimality criterion, a blend containing 0.50 fraction as 47h thermally treated WVO used for an engine torque of 14.72 Nm was identified as the optimum condition for minimum FC and maximum TE. The experimentally obtained FC and TE at factor setting nearest to the optimum recorded a value of 0.5094 kg/h and 23.70%, respectively. Real-life evaluation of diesel-waste frying oil (D-WFO) blend showed similar FC and TE as D-WVO blend at the optimized conditions. At 60% load (or 2.5 kW BP) in a CI engine, the proposed fuel blend was noted to result in better emissions, with 14.3% less CO and 17.6% fewer NO_x emissions as compared to conventional diesel. The present study demonstrated the potential of WVO, with varying thermal exposure time, as an alternate diesel fuel fraction.

5.5 References

[1] Mat, S. C., Idroas, M. Y., Hamid, M. F., Zainal, Z. A. (2018), Performance and emissions of straight vegetable oils and its blends as a fuel in diesel engine: A review", Renew. Sustain. Energy Rev., v. 82, pp. 808-823.

[2] Dabi, M., Saha, U. K. (2019), "Application potential of vegetable oils as alternative to diesel fuels in compression ignition engines: A review", J. Energy Institute, v. 92, pp. 1710-1726.

[3] Patel, P. D., Lakdawala, A., Patel, R. N. (2018), "Box–Behnken response surface methodology for optimization of operational parameters of compression ignition engine fueled with a blend of diesel, biodiesel and diethyl ether", Biofuels, v. 7, pp. 87-95.

[4] Yusri, I. M., Abdul Majeed, A. P. P., Mamata, R., Ghazali, M. F., Awad, O. I., Azmi, W. H. (2018), "A review on the application of response surface method and artificial neural network in engine performance and exhaust emissions characteristics in alternative fuel", Renew Sustain Energy Rev, v. 90, pp. 665–686.

[5] Singh, R. P., Debnath, S. (2011), "Heat and mass transfer in foods during deep-fat frying", In Kulp, K., Loewe, R., Lorenz, K., Gelroth, J. "Batters and breadings in food processing", Elsevier Inc. pp. 185–206.

[6] Sharoba, A. M., Ramadan, M. F. (2012), "Impact of Frying on Fatty Acid Profile and Rheological Behaviour of Some Vegetable Oils", J Food Process Technol, v. 1, pp. 3-7.

[7] Prabu, S. S., Asokan, M. A., Roy, R., Francis, S., Sreelekh, M. K. (2017), "Performance, Combustion and Emission Characteristics of Diesel Engine fueled with Waste Cooking Oil Bio-diesel or diesel blends with Additives", Energy, v. 122, pp. 638-648.

[8] Heywood, J. B. (1988), Internal Combustion Engine Fundamentals, McGraw-Hill Book Company, New York.

[9] Box, G. E.P., Draper, N. R. (1987), Empirical Model Building and Response Surfaces. John Wiley and Sons, NY, pp. 205–477.

[10] Myers, R. H., Montgomery, D. C., Anderson-Cook, C. M. (2016), Response Surface Methodology: Process and Product Optimization using Designed Experiments, 4th Edition, John Wiley & Sons, NJ.

[11] Majhi, S., Ray, S. (2016), "A study on production of biodiesel using a novel solid oxide catalyst derived from waste", Environ Sci Pollut Res, v. 23, pp. 9251-9259.

[12] Tripathy, D. B., Mishra, A. (2017), "Microwave Synthesis and Characterization of Waste Soybean Oil-Based Gemini Imidazolinium Surfactants with Carbonate Linkage", Surface Rev Letters, v. 24, pp. 1750062.

[13] Agarwal, A. K., Dhar, A. (2013), "Experimental investigations of performance, emission and combustion characteristics of Karanja oil blends fuelled DICI engine", Renew Energy v. 52, pp. 283-291.

[14] Nettles-Anderson, S., Olsen, D. B., Johnson, J. J., Enjalbert, J. N. (2014), "Performance of a Direct Injection Engine on SVO and Biodiesel from Multiple Feedstocks", J Power Energy Eng, v. 2, pp. 1-13.

[15] Asokan, M. A., Prabu, S. S., Kamesh, S., Khan, W. (2018), "Performance, combustion and emission characteristics of diesel engine fuelled with papaya and watermelon seed oil bio-diesel/ diesel blends", Energy, v. 145, pp. 238-245.

[16] Bupesh Raja, V. K., JayaPrabakar, J. (2017), "Performance and emission characteristics of cashew nut shell oil on CI engine", Int. J Ambient Energy, v. 40, pp. 563-565.

[17] Martin, M. L. J., Geo, V. E., Nagalingam, B. (2016), "Effect of fuel inlet temperature on cottonseed oil-diesel mixture composition and performance in a DI diesel engine", J Energy Institute, v. 90, pp. 563-573.

[18] Sonar, D., Soni, S. L., Sharma, D., Srivastava, A., Goyal, R. (2015), "Performance and emission characteristics of a diesel engine with varying injection pressure and fuelled with raw mahua oil (preheated and blends) and mahua oil methyl ester", Clean Technol Environ Policy, v. 17, pp. 1499–1511.

[19] Devan, P. K., Mahalakshmi, N. V. (2009), "Performance, emission and combustion characteristics of poon oil and its diesel blends in a DI diesel engine", Fuel, v. 88, pp. 861–867.

[20] Das, M., Sarkar, M., Datta, A., Santra, A. K. (2017), "An experimental study on the combustion, performance and emission characteristics of a diesel engine fuelled with diesel-castor oil biodiesel blends", Renew. Energy, v. 119, pp. 174-184.

[21] Sivaramakrishnan, K. (2017), "Investigation on performance and emission characteristics of a variable compression multi fuel engine fuelled with Karanja biodiesel–diesel blend", Egyptian J Petroleum, v. 27, pp. 127-136.

[22] Miraculas, G. A., Bose, N., Raj, R. E. (2016), "Optimization of Biofuel Blends and Compression Ratio of a Diesel Engine Fueled with *Calophyllum inophyllum* Oil Methyl Ester", Arabian J Sci Eng, v. 41, pp. 1723–1733.

[23] Murugapoopathi, S.; Vasudevan, D. (2019), "Performance, combustion and emission characteristics on VCR multi-fuel engine running on methyl esters of rubber seed oil", J Thermal Anal Calorimetry, v. 138, pp. 1329–1343.

[24] Dalta, R., Puli, R. K., Chandramohan, V. P., Geo, V. E. (2019), "Biodiesel Production Process, Optimization and Characterization of *Azadirachta indica* Biodiesel in a VCR Diesel Engine", Arabian J Sci Eng, v. 44, pp. 10141–10154.

[25] Emiroglu, A. O., Sen, M. (2018), "Combustion, performance and emission characteristics of various alcohol blends in a single cylinder diesel engine", Fuel, v. 212, pp. 34-40.

[26] Saxena, M. R., Maurya, R. K. (2017), "Optimization of engine operating conditions and investigation of nano-particle emissions from a non-road engine fueled with butanol/diesel blends", Biofuels.

[27] Yesilyurt, M. K. (2019), "The effects of the fuel injection pressure on the performance and emission characteristics of a diesel engine fueled with waste cooking oil biodiesel-diesel blends", Renew Energy, v. 132, pp. 649-666.

[28] Balakumar, R., Sriram, G, Arumugan, S. (2019), "Feasibility Study of Biodiesel Synthesis from Waste Ayurvedic Oil: Its Evaluation of Engine Performance, Emission and Combustion Characteristics", Arabian. J Sci Eng, v.. 44, pp. 8067–8079.

[29] Attia, A. M. A., Hassaneen, A. E. (2016), "Influence of diesel fuel blended with biodiesel produced from waste cooking oil on diesel engine performance", Fuel, v. 167, pp. 316-328.

[30] Al-Dawody, M., Jazie, A. A., Abbas, H. A. (2019), "Experimental and simulation study for the effect of waste cooking oil methyl ester blended with diesel fuel on the performance and emissions of diesel engine", Alexandria Eng J, v 58, pp. 9-17.

Chapter 6

TO DEFINE WASTE VEGETABLE OIL-BASED BLEND FUELS FOR UTILIZATION IN A CI ENGINE

6.1 Introduction

The vegetable oil has been reportedly used in CI engines as an alternative fuel by blending with diesel fuel and their performance and emission analyses were studied by many researchers [1]. But the growing concern based on the food versus fuel dispute has shifted the recent focus towards the exploration of other renewable alternatives as fuel or fuel blends in CI engines [2]. In this regard, WVO from cooking or frying, if utilized judiciously, can be of great value. However, limited literature is available on the evaluation of unmodified WVO as blend fuel with fossil fuel for emission characterization and performance analysis on utilization in a CI engine [3]. Hence, a research scope exists in valorizing WVO by blending with fossil fuels and utilizing it as fuel in a CI engine to offer a sustainable and eco-friendly solution to the ongoing fuel crisis and environmental disposal issue. WVO, upon preheating or chemical modifications, was reported to be usually blended with diesel [4-6]. Also, kerosene blending with vegetable oils was found to lower their impediments and provide engine performance similar to diesel [7-9].

Thus, this chapter provides an amalgamation of two separate studies on the variations in the fuel properties and engine performance and emission characteristics of a CI research engine fueled separated with diesel-WVO (D-WVO) and kerosene-WVO (K-WVO) blends.

6.2 Methodology

6.2.1 Materials

The diesel was sourced from a local diesel vehicle refueling station (Agartala, TR, India). Kerosene was purchased from the local market (Agartala, TR, India). The vegetable oil was sourced from the brands of edible vegetable oil available in the local market (Agartala, TR, India).

6.2.2 Preparation of different WVO blends

The volumetric blends of WVO with diesel and kerosene were prepared in three different blend proportions, namely, 0.20, 0.35, and 0.50 fractions as WVO. WVO blend fractions above 0.50 were not considered as conventional diesel engines, fitted with conventional fuel injectors, have not been reported to accommodate higher WVO fractions in blend fuel [6]. Thus, the different blends evaluated in the study were: D-WVO_0.20 (0.80 fraction diesel + 0.20 fraction WVO), D-WVO_0.35 (0.65

fraction diesel + 0.35 fraction WVO), D-WVO_0.50 (0.50 fraction diesel + 0.50 fraction WVO), K-WVO_0.20 (0.80 fraction kerosene + 0.20 fraction WVO), K-WVO_0.35 (0.65 fraction kerosene + 0.35 fraction WVO), and K-WVO_0.50 (0.50 fraction kerosene + 0.50 fraction WVO).

The D-WVO blends containing WVO of three thermal treatment durations (24H, 48H, and 72H) are assessed in the first study. The three WVO stocks are already discussed in Chapter 5, where an optimization study on the fuel consumption and thermal efficiency of a CI engine is presented. However, full-scale engine performance and emission study are eminent to showcase the applicability of D-WVO blends in a CI engine.

6.2.3 Characterization of test fuels

The various WVO blends were characterized using ASTM standard test methods and compared with ASTM D975 and IS 1460 diesel standards, as mentioned in Section 3.2.3 in Chapter 3.

6.2.4 Assessment of engine performance and emission characteristics

The test engine setup was the same as employed for the study presented in Section 3.2.4 in Chapter 3.

6.2.4.1 Uncertainty analysis

Table 6.1 Uncertainty of various measurements throughout the study.

Measurement	Accuracy	% Uncertainty	Measurement technique
Load	± 0.1 kg	± 0.2	Strain gauge-type load cell
Engine speed	± 10 rpm	± 0.1	Magnetic pickup type
Fuel flow	± 0.1 cu. cm	± 1.0	Volumetric measurement
Time	± 0.01 sec	± 0.2	Digital stopwatch
Carbon monoxide	± 0.01 % vol.	± 0.2	NDIR principle
Carbon dioxide	± 0.1 % vol.	± 0.15	NDIR principle
Hydrocarbon	± 10 ppm	± 0.2	NDIR principle
Nitrogen oxide	± 20 ppm	± 0.2	Electrochemical measurement

The accuracy of results is determined using uncertainty analysis. The errors in the measured values define the uncertainty associated with the measurement. The uncertainties of various measurements used were computed as per the method reported in the literature [10] and tabulated in Table 6.1. All the experiments were carried out in triplicate to minimize the errors. The overall uncertainty of the measurements was computed to be less than 5.0 %.

6.3 Experimental findings

6.3.1 Fuel characterization

6.3.1.1 Effect of heating duration on WVO properties

The characteristics of WVO subjected to different heating duration are significantly different. The increase in heating duration showed a slight shift and a substantial increase in the UV absorbance peak of WVO between 380 to 460 nm (Fig. 6.1(a)). Straight vegetable oil (WVO_0H) was considerd as the baseline for the assessment. The increase in absorbance can be linked with the increased formation of oxidized conjugated and polymerized products in WVO [11]. Moreover, with the increase in heating duration, a decrease in the degree of unsaturation of the oil was observed (Fig. 6.1(b)). The ratio of the absorbance at 3005 cm^{-1} (stretching vibrations of cis C=C bonds) and 2850 cm^{-1} (symmetric stretching vibrations of aliphatic -CH bonds) in the FT-IR spectra can be used to determine the degree of unsaturation in the oil [12]. The values of the degree of unsaturation were 0.25 for WVO_24H and nearly 0.11 for both WVO_48H and WVO_72H. The loss in the unsaturation of the oil was 56% when heat treatment was continued for 48 hours. Thus, elongated heating can change the physicochemical properties of the oil. A significant change was noted in the physical properties of WVO samples. An increase in heating duration caused an increase in the density and kinematic viscosity (Fig. 6.1(c)). In comparison to diesel, the viscosity of the WVO specimens was noted to be significantly higher. Since, engine performance of fuel relies on a fine balance between density and viscosity of the fuel, so different WVO in various blend proportions with diesel was expected to affect the droplet atomization and penetration of the fuel jet in the engine [13].

Accordingly, the evaluation of different WVO in various blend proportions as proposed is of logical significance.

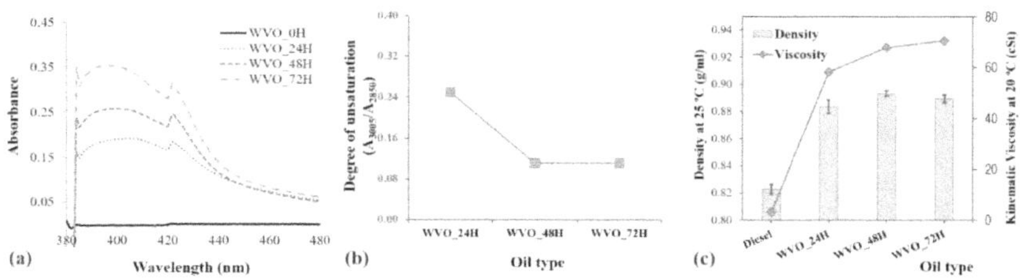

Fig. 6.1 Effect of heating duration on WVO properties - **(a)** UV-Vis absorbance spectra, **(b)** Degree of unsaturation from FT-IR spectra, and **(c)** Density and kinematic viscosity.

6.3.1.2 Properties of the blend fuels

The WVO evaluated in the present study was found to have good miscibility with diesel fuel. Due to the high viscosity of WVO compared to diesel, WVO blend proportion beyond 0.50 was not considered for the test CI engine fitted with a conventional fuel injection system. A comparison of properties of diesel fuel, blended diesel fuel with different WVO fractions are presented in Table 6.2 for 48H WVO. Since similar trends were also noted for the diesel blends with 24H and 72H WVO, hence additional values were not tabulated. The density, viscosity as well as flash and fire points for the blended fuel increased as the fraction of WVO was increased in the blends. The rise in the viscosity of blends with increasing WVO fraction is noticeable. The flashpoints, fire points, and heating values were comparable to that of diesel fuel. An increase in flash and fire points and a decrease in the calorific value of the blend were recorded for increasing WVO fractions. The density of all the blends falls in the range specified for diesel fuel (0.82 – 0.86 g/cc), but the viscosities exceeded the standard range for diesel (IS 1460). The cloud and pour points of the WVO blends are low, which shows that such blends shouldn't be used in colder climates where a severe drop in outside temperature may cause flow impediments for the WVO blends.

Table 6.2 Properties of the various fuel stocks.

Fuel samples	Density at 15°C (g/cc)	Viscosity at 40°C (mm²/s)	Flash point (°C)	Fire point (°C)	Cloud point (°C)	Pour point (°C)	Oxidation stability (h)	Cetane Index	Calorific value (kJ/kg)
Diesel	0.823	3.12	49	58	1	-17	>20	52.25	46,646
Kerosene	0.797	1.68	46	54	-32[a]	-47[a]	>20	43.16	46,974
WVO-24H	0.876	30.46	178	>200	6	3	12.78[c]	46.09	39,665
WVO-48H	0.881	31.58	189	>200	6	3	12.89[c]	46.28	39,990
WVO-72H	0.887	33.74	203	>200	6	3	12.91[c]	46.31	40,126
D-WVO_0.20	0.841	6.01	53	70	1	-9	-----	51.27	45,321
D-WVO_0.35	0.851	9.36	56	76	4	-5	-----	50.16	44,372
D-WVO_0.50	0.861	15.24	61	89	9	-2	-----	49.67	43,208
K-WVO_0.20	0.812	5.59	52	69	-5	-16	-----	45.14	45.577
K-WVO_0.35	0.827	6.72	55	74	1	-8	-----	45.92	44.529
K-WVO_0.50	0.844	8.98	59	85	3	-2	-----	46.11	43.842
Standard range (IS 1460:2017)	0.810 - 0.845	2.0 – 4.5	35 (min)	-----	3 (max)[b]	-----	20 (min)	46 (min)	43,000 (min)
Test method	ASTM D 1217	ASTM D2983	ASTM D93		ASTM D2500		-----	ASTM D976	ASTM D240
Instruments	Pycno-meter	Digital rotational viscometer	Pensky Martens closed cup apparatus		Cloud point & pour point apparatus		-----	Atmospheric distillation apparatus	Digital bomb calorimeter

[a] – value reported in the literature.
[b] – Diesel standard value for winter season in India.
[c] – Oxidation stability = 22.318 – 0.234*(linoleic acid wt%) (method proposed by Ref [14]).

6.3.2 Engine performance and emission characteristics of D-WVO blends

6.3.2.1 Engine performance characteristics of D-WVO blends

The engine performance was evaluated for brake-specific fuel efficiency (BFSC) and brake thermal efficiency (BTE) at three different load settings, viz. 20%, 40%, and 60% of the maximum load. BSFC is a very important parameter for studying the performance of fuel in an engine and defines the fuel consumption for a unit of engine power. An increase in the amount of fuel necessary to produce a specific engine output power increases the BSFC [15]. The BSFC of the WVO-diesel blends were very much comparable with diesel at various levels of engine load (Fig. 6.2). The BTE suggests how efficiently the heat produced due to the combustion of fuel in the engine is converted to mechanical work [16, 17]. BTE was noted to increase relative to load irrespective of the blends. This phenomenon was due to the increment in power developed and decrement in heat loss upon the increase in engine load [15]. At lower load, BTEs for all the diesel blends were at par with diesel fuel (Fig. 6.3). Since the output power generated by the engine upon the combustion of a specific fuel at a particular engine load is comparable for all the WVO-diesel blends of different proportions, so the BSFC and BTE of WVO-blends were similar to that of diesel fuel. WVO has a lower heating value but higher viscosity and density compared to diesel (Table 6.2). Increased WVO fraction in the blend fuel resulted in marginally reduced calorific value and volatility but increased lubricity. Higher viscosity minimizes the backflow across the piston clearance of the injection pump and higher lubricity reduces the frictional losses [18]. Thus, the BSFC values of the WVO-blends were similar to that of diesel fuel. The presence of oxygen and the good lubricating property of WVO resulted in better combustion and lower frictional losses. Similar explanations are reported in the literature [19, 20]. Thus, marginally lower calorific values of D-WVO blends compared to diesel are compensated and diesel blends with 0.20 fraction of WVO showed higher BTE (for all heating duration of WVO) in comparison to diesel.

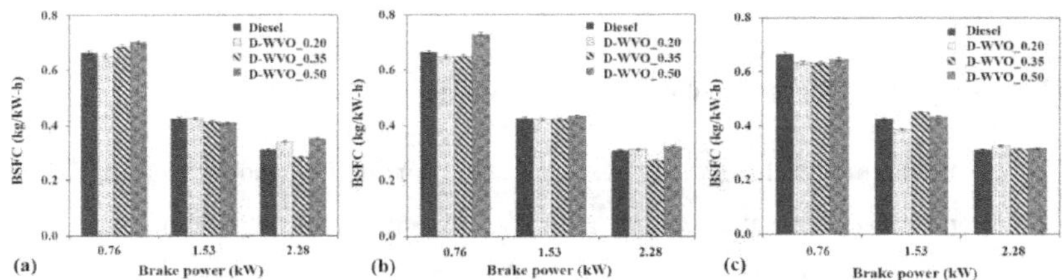

Fig. 6.2 BSFC of the engine for different diesel blends at varying engine loads. **(a)** Blends with 24H WVO fractions, **(b)** blends with 48H WVO fractions, and **(c)** blends with 72H WVO fractions.

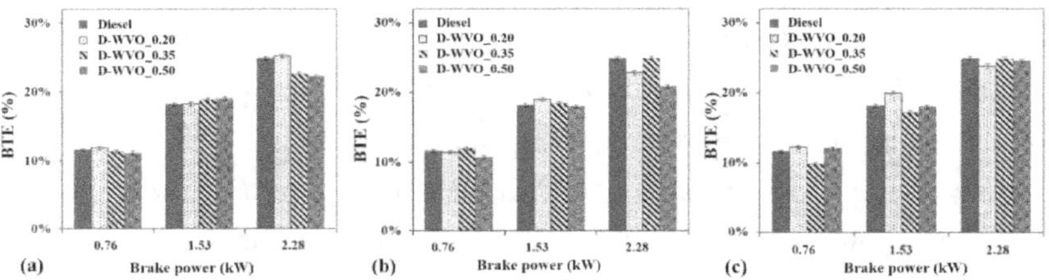

Fig. 6.3 BTE of the engine for different diesel blends at varying engine loads. **(a)** Blends with 24H WVO fractions, **(b)** blends with 48H WVO fractions, and **(c)** blends with 72H WVO fractions.

6.3.2.2 Engine emission characteristics of D-WVO blends

The engine emission characteristics were investigated for carbon monoxide (CO), unburnt hydrocarbon (HC), and the oxide of nitrogen (NO_x) emissions at four different load settings, viz. 0%, 20%, 40% and 60% of the maximum load.

The CO emission occurs due to the incomplete combustion of fuel inside the engine cylinder. At the no-load condition (0% load), the temperature of the gas inside the engine cylinder remains low which results in incomplete combustion in the gas phase and relatively high CO emission. Thus, CO emission for all the blends of WVO-diesel was comparable with diesel and emission values decreased with increased engine load (Fig. 6.4). Higher oxygen content in fuels of bio-origin (like WVO) leads to higher combustion temperature and resulted in more conversion of CO to CO_2.

Accordingly, at high engine loads, the resultant CO emission in the exhaust was lower for all the WVO-diesel blends than diesel fuel. Similar results were also reported in the literature [15, 19].

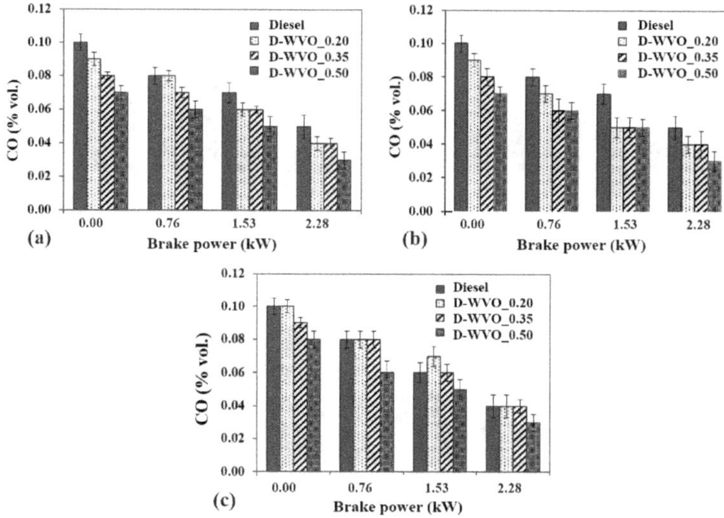

Fig. 6.4 CO emission from the engine for different diesel blends at varying engine loads. **(a)** Blends with 24H WVO fractions, **(b)** blends with 48H WVO fractions, and **(c)** blends with 72H WVO fractions.

The emission of HC was observed to increase monotonically with the increase in load (Fig. 6.5). Hydrocarbon (HC) emission from the engine also takes place due to the incomplete combustion of fuel vapor in the engine cylinder [20]. This can be attributed to the increased quantity of fuel injected at higher loads. Higher HC emission for blend fuels compared to diesel was recorded at lower loads. Lower engine temperatures at the no-load (0% load) condition may be attributed to incomplete combustion of denser WVO blends and higher HC emission compared to diesel. However, higher oxygen content in WVO blends resulted in better combustion at higher engine load (at higher engine temperature). Accordingly, all WVO-diesel blends produced lesser unburnt HC emissions than unblended diesel.

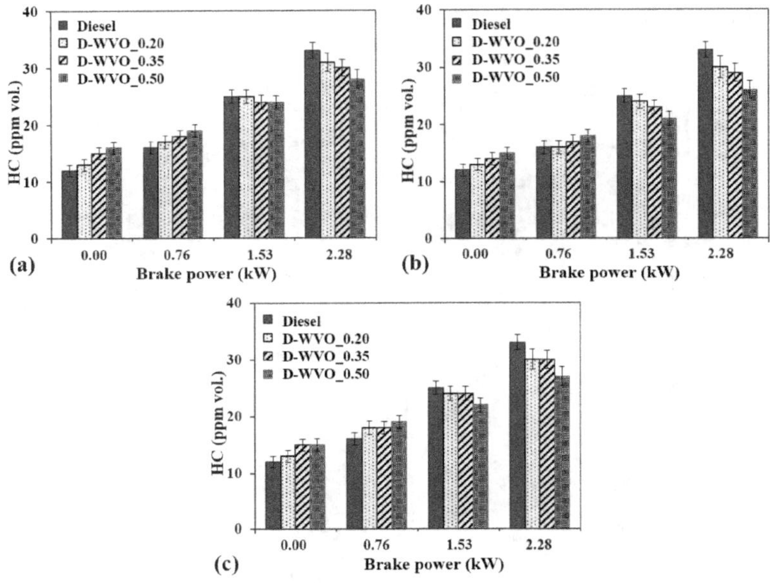

Fig. 6.5 HC emission from the engine for different diesel blends at varying engine loads. **(a)** Blends with 24H WVO fractions, **(b)** blends with 48H WVO fractions, and **(c)** blends with 72H WVO fractions.

The oxides of nitrogen (NO_x) are major pollutants emitted from CI engines due to high in-cylinder temperature and non-premixed combustion [20]. NO_x emission is the most important emission characteristic of vegetable oil [21]. Due to enhanced combustion and higher in-cylinder temperature at higher loads, as compared to lower loads, there was an increase in the heat release rate and combustion. Thus, NO_x release was found to increase with an increase in engine load. A similar trend was noted in this study (Fig. 6.6). At no load (0% load) conditions, the WVO-diesel blend produced lower NO_x emissions than diesel fuel. But, at increased load (20% and 40% load), higher NO_x emission was noted for all WVO blends as compared to diesel. At the highest load condition (60% load), the NO_x emissions for all the blends were lower than those for diesel. Higher oxygen content in WVO blends lowers the requirement of oxygen from the intake air for combustion of the WVO blends which results in the intake air diluting the product gas reduce NO_x formation [20].

Accordingly, slightly lower NO$_x$ emission was noted for all WVO blends at the highest (60% load) load condition as compared to diesel.

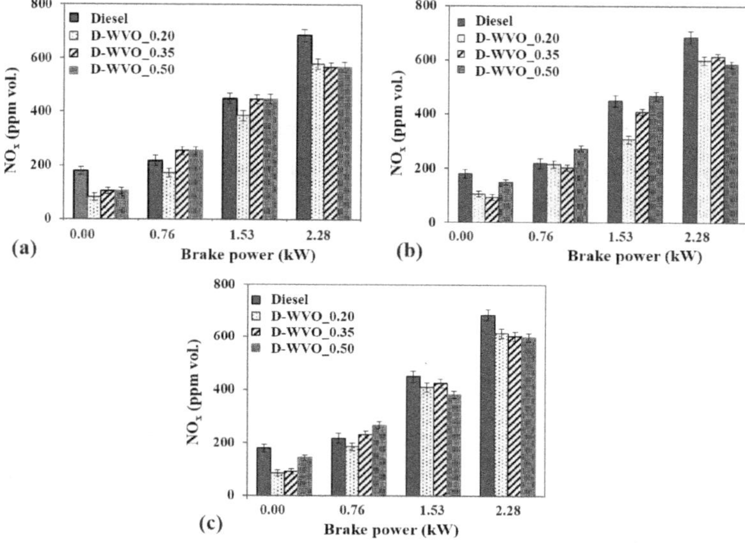

Fig. 6.6 NO$_x$ emission from the engine for different diesel blends at varying engine loads. **(a)** Blends with 24H WVO fractions, **(b)** blends with 48H WVO fractions, and **(c)** blends with 72H WVO fractions.

6.3.3 Engine performance and emission characteristics of K-WVO blends

6.3.2.1 Engine performance characteristics of K-WVO blends

The engine performance parameters evaluated to determine the performance of the fuel during engine operation include brake specific fuel consumption (BSFC) and brake thermal efficiency (BTE). The K-WVO blends were assessed for engine performance parameters at three different engine load settings and compared against diesel fuel.

The BSFC of various fuels under different load conditions is presented in Fig. 6.7(a). An increase in engine load causes a higher mass flow of fuel and facilitates better combustion. Accordingly, decreased BSFC was noted for increased brake power (engine load). The study reported in the literature concurred with the reported trend of lower BSFC at higher load settings [22]. At medium brake power (medium load), the BSFCs of K-WVO blends were significantly lower than diesel fuel. Higher volatility of kerosene plays a positive role in generating an appropriate air-fuel mixture that favors better combustion. Such an explanation is presented by researchers who have worked with vegetable oil blends in kerosene [9]. Hence, the lowest BSFC was noted for K-WVO_0.20 among the K-WVO blends at all engine loads. Higher density and lesser calorific value of WVO compared to diesel and kerosene resulted in increased fuel consumption and slightly higher BSFC of K-WVO_0.50 blend (0.321 kg/kWh), relative to diesel (0.333 kg/kWh). The BSFCs of all K-WVO blends were found analogous to diesel at all load settings.

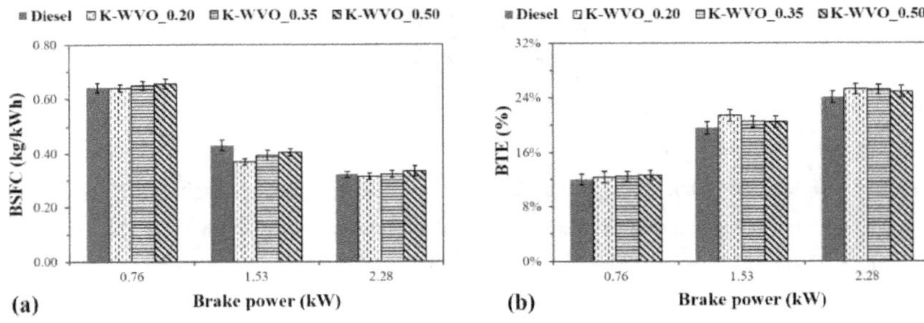

Fig. 6.7 Variations in engine performance of various K-WVO blends – **(a)** BSFC and **(b)** BTE.

As per the literature, the properties of fuel influence the BTE of a CI engine [18, 22]. Thus, the BTEs of various fuels were determined at different engine load settings (Fig. 6.7(b)). Higher BTE was noted at increased engine load for the various test fuels. Increased BTE can be linked with reduced BSFC and increased output power at a particular load setting, as evident from the literature [23]. Thus, the blend containing a higher kerosene fraction (K-WVO_0.20) recorded the highest BTE of 25.22% among the fuels evaluated. The BTE of K-WVO_0.35, K-WVO_0.50, and diesel fuel were recorded as 25.16%, 24.84%, and 24.06%, respectively. As BSFC of K-WVO blends were comparable to diesel, so all the K-WVO blends demonstrated BTE comparable to diesel fuel at low engine load settings. Due to increased output power at analogous BSFC, the BTE values of the K-WVO blends were marginally better (3.3 - 4.8%) than diesel at the higher load setting of the engine.

WVO in K-WVO blend reduced frictional loss due to higher lubricity and favored improved combustion due to higher oxygen content. Thus, the lower lubricity of kerosene was compensated by the WVO in the blend. Similarly, higher density and viscosity that impede the atomization and fine droplet formation were offset by more volatile kerosene fractions in the K-WVO blend. This fine balancing of attributes resulted in similar engine performance of the K-WVO blends in comparison to diesel at most load conditions. The observation aligned well with the trend of the properties of the blend fuels, wherein the properties of blend fuels and diesel were noted as comparable.

6.3.2.2 Engine emission characteristics of K-WVO blends

The characteristics of emission during engine operation have been known to depend on the fuel properties and engine performance characteristics [18]. Accordingly, the emission characteristics for various fuels considered in the study were determined from engine operation. Carbon monoxide (CO) and unburnt hydrocarbon (HC) are the main products of fuel combustion in a CI engine [22]. The brake-specific emissions of carbon monoxide (CO), unburnt hydrocarbon (HC), and oxides of nitrogen (NO_x) were determined to characterize the emission characteristics of various test fuels. The specific emission profiles of the fuels for various load conditions are presented in Fig. 6.8.

The limited combustion of fuel results in the emission of CO from the CI engines [18]. Fig. 6.8(a) illustrates the variations of CO emission for various test fuels. In agreement with the literature, the CO emission decreased with increased engine load [24]. Higher oxygen content WVO favors complete combustion and lowers CO emission [23]. Thus, increased WVO fraction enhanced the oxygen content in blend fuels and favored complete combustion that caused lower CO emission for the K-WVO blends in comparison to diesel. The brake-specific CO emission was the lowest for the K-WVO blend with the highest WVO fraction (K-WVO_0.50). The brake-specific CO emission of K-WVO_0.50 was 44.1% lower than diesel.

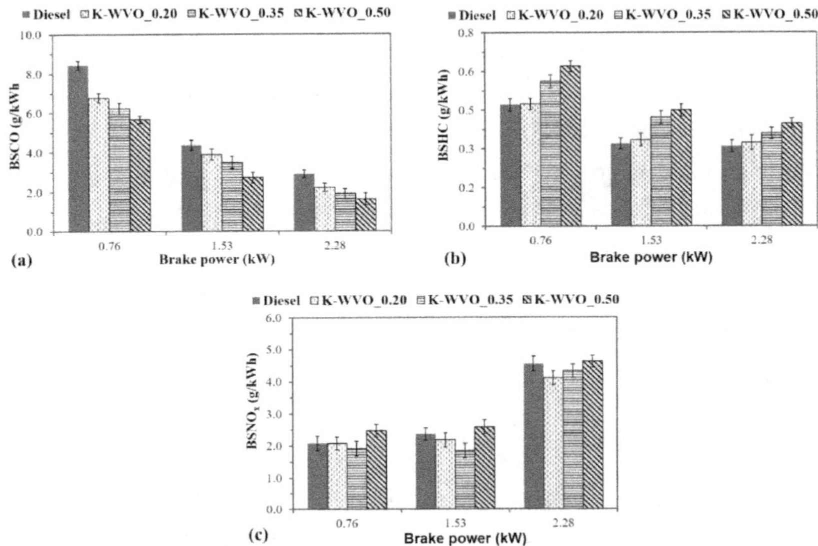

Fig. 6.8 Variations in engine emissions of various K-WVO blends – **(a)** CO, **(b)** HC, and **(c)** NO$_x$.

The poor atomization or spray properties of the fuel are responsible for the rise in HC emission [18, 22]. Fig. 6.8(b) presents the brake-specific HC emission of the test fuels under various engine load conditions. Increased HC emission was noted for all K-WVO blends relative to diesel. The higher proportion of WVO in the blend increased the HC formation under all load conditions. Increased density and viscosity of WVO lead to improper mixing, poor atomization which lead to the formation of blowout region during the expansion stroke in the far end of the combustion chamber and hence resulted in higher HC emissions. A similar trend was observed by researchers who have blended vegetable oils and biodiesel with kerosene [8, 9]. Thus increased WVO in blend recorded higher HC emission compared to diesel. Lower in-cylinder temperature for low load setting may have been the reason behind a relatively higher HC emission than diesel. At high load conditions, in-cylinder temperature increases as a result of the unfavorable impact of higher density and viscosity of WVO on the atomization of the fuel is reduced and the difference between HC emissions of the blend relative to diesel was noted to lessen. Thus, at the high engine load condition, the relative rise in specific HC emission for K-WVO blends was less than the trends observed at low load conditions. A similar observation of reduced HC emission for increased in-cylinder temperature was also reported in the literature [25].

Literature suggests that the formation of NO_x in engine operation is linked with higher cylinder temperature during fuel combustion [18]. The variations in NO_x emission for various test fuels are shown in Fig. 6.8(c). An increase in engine load increases the cylinder temperature. Accordingly, NO_x emission was found to increase with increased load conditions. Also, higher oxygen content in WVO tends to raise the in-cylinder temperature during combustion in an engine operation, as inferred in the literature [23]. Thus, higher NO_x emissions for K-WVO blends with higher WVO content. Since the fuel properties and performance parameters of engine operation were comparable between diesel and K-WVO blends, so the emissions of NO_x for most K-WVO blends were analogous to that of diesel under various load conditions. The K-WVO_0.50 blend recorded marginally higher NO_x emission relative to diesel at low engine load.

The analysis of emission characteristics corroborated the findings of fuel property and engine performance comparison between K-WVO blends and diesel. The emission comparison established that most K-WVO blends were nearly equivalent to diesel and can be expected to successfully substitute diesel as fuel. The replacement of diesel by K-WVO blends is a novel approach in alternative fuel research as a similar scheme was never adopted and reported earlier in the literature. Thus, the combined study on fueling a CI research engine with D-WVO and K-WVO blends has proved that such blends have the potential to run light-duty CI engines without any operational issue. Moreover, a general reduction in CO and NO_x emission, while using WVO blend fuels is a promising outcome of the studies.

6.4 Summary

The direct blending of WVO is proposed as an approach towards waste valorization and alternative fuel research. WVO is found to form stable bends with both diesel and kerosene. Elevated BSFC and slightly lower BTE of WVO blends are obvious because of relatively higher densities and viscosities. However, engine operations are expected to go smoothly for all the blends. Moreover, better emission profiles linked with such blends show that such blend fuels are environmentally better. Moreover, the economic benefits associated with WVO blends may incentivize the concept of utilizing WVO as alternative fuel fractions in existing CI engines. From this study, it is confirmed that a problematic waste like WVO, if used judiciously by blending with diesel or kerosene, can prove to be a better alternative to diesel fuel.

6.5 References

[1] Mat, S. C., Idroas, M. Y., Hamid, M. F., Zainal, Z. A. (2018), Performance and emissions of straight vegetable oils and its blends as a fuel in diesel engine: A review", Renew. Sustain. Energy Rev., v. 82 pp. 808-823.

[2] Dabi, M., Saha, U. K. (2019), "Application potential of vegetable oils as alternative to diesel fuels in compression ignition engines: A review", J. Energy Institute, v. 92, pp. 1710-1726.

[3] Capuano, D., Costa, M., Di Fraia, S., Massarotti, N., Vanoli, L. (2017), "Direct use of waste vegetable oil in internal combustion engines", Renew Sustain Energy Rev, v. 69, pp. 759-770.

[4] Kalam, M. A., Masjuki, H. H., Jayed, M. H., Liaquat, A. M. (2011), "Emission and performance characteristics of an indirect ignition diesel engine fueled with waste cooking oil", Energy, v. 36, pp. 397–402.

[5] Senthil Kumar, M., Jaykumar, M. A. (2014), "Comprehensive study on performance, emission and combustion behavior of a compression ignition engine fueled with WCO (waste cooking oil) emulsion as fuel", J Energy Institute, v. 87, pp. 263–271.

[6] Dhanasekaran, R., Krishnamoorthy, V., Rana, D., Saravanan, S., Nagendran, A., Rajesh Kumar, B. (2017), "A sustainable and eco-friendly fueling approach for direct-injection diesel engines using restaurant yellow grease and n-pentanol in blends with diesel fuel", Fuel, v. 193, pp. 419–431.

[7] Ravindra, M., Aruna, Vardhan, H. (2017), "Investigation on the performance of a variable compression ratio engine operated with raw cardanol kerosene blends", Biofuels, pp. 1-7.

[8] Aydın H (2016), "Scrutinizing the combustion, performance and emissions of safflower biodiesel–kerosene fueled diesel engine used as power source for a generator", Energy Convers Manag, v. 117, pp. 400-409.

[9] Bayindir, H., Isık, M. K., Argunhan, Z., Yücel, H. L., Aydın, H. (2017), "Combustion, performance and emissions of a diesel power generator fueled with biodiesel-kerosene and biodiesel-kerosene-diesel blends", Energy, v. 123, pp. 241-251.

[10] Holman, J. P. (2004), "Experimental techniques for engineers", 1st ed., Tata McGraw Hill, New Delhi.

[11] Majhi, S., Ray, S. (2016), "A study on production of biodiesel using a novel solid oxide catalyst derived from waste", Environ Sci Pollut Res, v. 23, pp. 9251-9259.

[12] Valdes, A. F., Garcia, A. B. (2006), "A study of the evolution of the physicochemical and structural characteristics of olive and sunflower oils after heating at frying temperatures", Food Chemistry, v. 98, pp. 214–219.

[13] Agarwal, D., Kumar, L., Agarwal, A. K. (2008), "Performance evaluation of a vegetable oil fueled compression ignition engine", Renew Energy, v. 33, pp. 1147–1156.

[14] Kumar, N. (2017), "Oxidative stability of biodiesel: Causes, effects and prevention", Fuel, v. 190, pp. 328-350.

[15] Attia, A. M. A, Hassaneen, A. E. (2015), "Influence of diesel fuel blended with biodiesel produced from waste cooking oil on diesel engine performance", Fuel, v. 167, pp. 316-328.

[16] Asokan, M. A., Prabu, S. S., Kamesh, S., Khan, W. (2018), "Performance, combustion and emission characteristics of diesel engine fuelled with papaya and watermelon seed oil bio-diesel/diesel blends", Energy, v. 145, pp. 238-245.

[17] Prabu, S. S., Asokan, M. A., Roy, R., Francis, S., Sreelekh, M. K. (2017), "Performance, Combustion and Emission Characteristics of Diesel Engine fueled with Waste Cooking Oil Bio-diesel or diesel blends with Additives", Energy, v. 122, pp. 638-648.

[18] Hazar, H., Aydin, H. (2010), "Performance and emission evaluation of a CI engine fueled with preheated raw rapeseed oil (RRO)–diesel blends", Applied Energy, v. 87, pp. 786–790.

[19] Agarwal, A. K., Dhar, A. (2013), "Experimental investigations of performance, emission and combustion characteristics of Karanja oil blends fuelled DICI engine", Renew Energy, v. 52, pp. 283-291.

[20] Das, M., Sarkar, M., Datta, A., Santra, A. K. (2017), "An experimental study on the combustion, performance and emission characteristics of a diesel engine fuelled with diesel-castor oil biodiesel blends", Renew Energy, v. 119, pp. 174-184.

[21] Murugesan, A., Umarani, C., Subramanian, R. Nedunchezhian, N. (2009), "Bio-diesel as an alternative fuel for diesel engines- A review", Renew Sustain Energy Rev, v. 13, pp. 653–662.

[22] Huang, H., Teng, W., Liu, O., Zhou, C., Wang, Q., Wang, X. (2016), "Combustion, performance and emission characteristics of a diesel engine under low-temperature combustion of pine oil–diesel blends", Energy Convers Manag, v. 128, pp. 317-326.

[23] Yesilyurt, M. K. (2019), "The effects of the fuel injection pressure on the performance and emission characteristics of a diesel engine fueled with waste cooking oil biodiesel-diesel blends", Renew Energy, v. 132, pp. 649-666.

[24] Sonar, D., Soni, S. L., Sharma, D., Srivastava, A., Goyal, R. (2015), "Performance and emission characteristics of a diesel engine with varying injection pressure and fuelled with raw mahua oil (preheated and blends) and mahua oil methyl ester", Clean Technol Environ Policy, v. 17, pp. 1499-1511.

Chapter 7

TO COMPARATIVELY ANALYZE WASTE VEGETABLE OIL BLENDS WITH DIESEL AND KEROSENE

7.1 Introduction

From the last two chapters, it became clear that WVO blends very well with both diesel and kerosene and the WVO blends show properties close to the diesel standards. The studies on engine performance and emission characteristics of WVO-diesel (D-WVO) blends and WVO-kerosene (K-WVO) blends proved that such blends can be easily used to operate existing CI engines. However, a question still remains – which fossil fuel is a better match for WVO blends? Thus, to answer this question, a comparative analysis of D-WVO and K-WVO blends was performed. Engine combustion, performance, emissions, and running cost, as well as potential environmental impacts of combustion of the WVO blends were considered as the parameters for this comparison.

The literature survey suggested that WVO has been used as a diesel blend, either upon preheating or with an additive [1]. Thus, it is pertinent to study the impact of the direct blending of WVO with diesel on fuel properties and also on the combustion, performance, and emission characteristics of a CI engine. Also, kerosene has been reported to lower the density and viscosity of vegetable oil blends and assists in improving engine performance [2, 3]. Thus, engine combustion, performance, and emissions of WVO blends containing kerosene need to be looked into.

The present study provides an overall comparison between D-WVO and K-WVO blends in terms of fuel properties and engine performance, combustion, and emission characteristics. Also, the economics of using the D-WVO and K-WVO blends were assessed. Finally, various potential environmental impacts associated with the combustion of WVO blend fuels were also analyzed using an open-ended life cycle assessment approach. Such a holistic comparison between D-WVO and K-WVO blends is a novel study that was never attempted to date.

7.2. Methodology

7.2.1 Preparation of WVO blends

The various WVO blends were prepared as discussed in Section 6.2.2 of Chapter 6. For comparison purposes, 0.20, 0.35, and 0.50 fractions containing WVO blends in diesel and kerosene were considered for the study.

7.2.2 Characterization of WVO blends

The various WVO blends were characterized using ASTM standard test methods and compared with ASTM D975 diesel standards, as mentioned in Section 3.2.3 in Chapter 3.

7.2.3 Assessment of engine characteristics

A 3.5 kW stationary direct injection, single-cylinder, compression ignition (CI) diesel engine (Model - TV1; Make: Kirloskar Oil Engines Ltd.) was used to conduct the engine test runs of various fuels. A water-cooled eddy-current dynamometer (Model – AJ10; Make: SAJ Test Plant Pvt. Ltd.) with a digital load cell (Model – 60001; Sensortronics Pvt. Ltd.) was paired with the engine. A five gas exhaust analyzer (Model - Digas 444; Make: AVL India Ltd.) was used to collect the emission data from the engine exhaust pipe. The specifications of the components of the experimental setup are presented in Table 7.1. The engine RPM sensor, piezoelectric pressure transducer, temperature sensors (K-type thermocouple), crank angle encoder, and orifice meter were all calibrated before the start of each test cycle. The sensor inputs were collected using a data acquisition system (16-bit, NI USB 6210) linked with an engine performance combustion analysis software (EngineSoft v9.01).

Table 7.1 Specifications of the test engine and various measuring units.

Parameters	Specifications
Engine	Kirloskar Oil Engines, Model – TV1
Strokes/ Number of cylinders	Four strokes/ One cylinder
Bore and stroke	87.5 mm and 110 mm
Compression ratio	17.5: 1 (fixed)
Swept volume	0.661 liters
Air intake	Natural aspiration
Rated power	3.5 kW @ 1500 RPM
Injection strategy	Direct injection
Injection pressure	210 bar
Injection timing	23° before the top dead center
Cooling system	Cooling water
Dynamometer	SAJ Test Plant, Model – AJ10
Load sensor	Sensotronics Sanmar, Model - 60001
Crank angle encoder	Kubler, Model - 8.3700.1321.0360
Pressure sensor	PCB Piezotronics, Model – HSM111A22
Temperature sensor	Radix, Model – PT100 (K-type)
Fuel flow transmitter	Yokogawa, Model - EJA110A-DMS5A-92NN
Airflow transmitter	Wika, Model – SL10
Exhaust gas analyzer	AVL Gas analyzer, Model - Digas 444
Resolution	O_2 and CO = 0.01% vol; CO_2 = 0.1% vol; HC and NO_x = 1 ppm

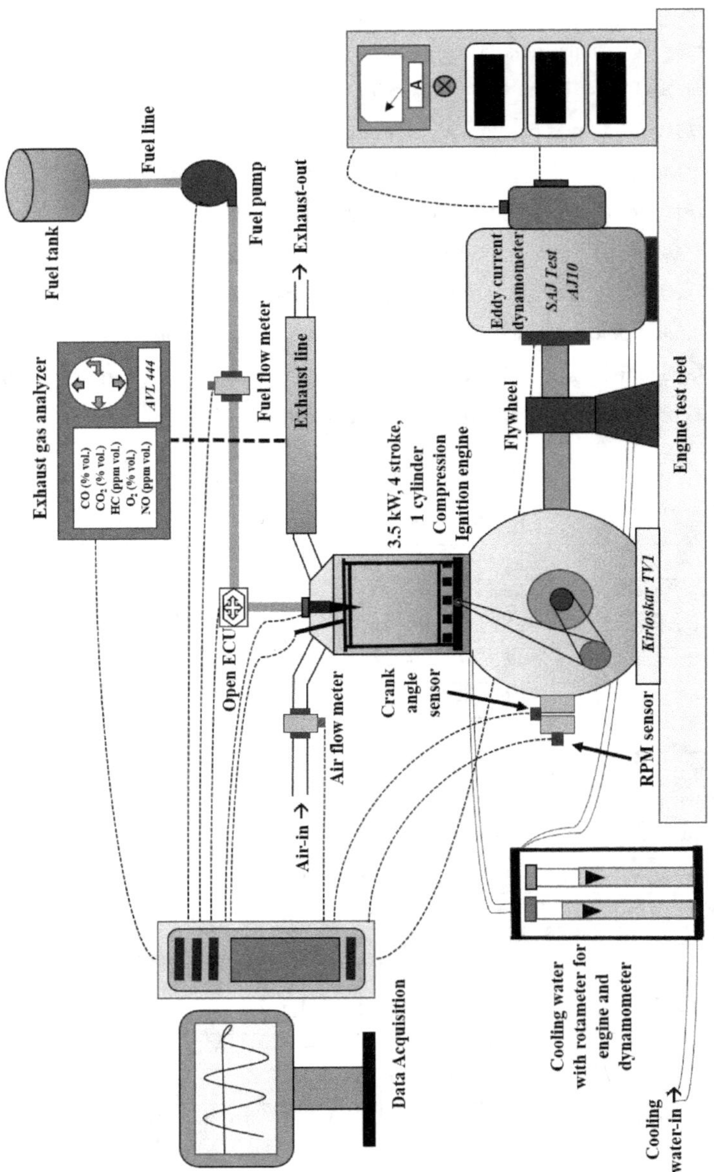

Fig. 7.1 CI engine test setup for the present study.

The engine operation was performed at a constant speed of 1500 RPM. The engine was warmed up under the no-load condition with a continuous flow of cooling water before every test runs to minimize variations in the operational condition. The fuels were tested at three load conditions between 0 to 60% (20%, 40%, and 60%) of the full load during engine operation. Due to operational issues and limitations, the engine operation could not be performed at a 100% load condition. Appropriate lubrication and engine cooling (using cooling water) were maintained to ensure the proper operational condition of the engine. After each run, the fuel line was flushed with diesel to avoid possible contamination among different fuels under evaluation. The uncertainties associated with various measurements are presented in Table 7.2.

Table 7.2 Uncertainties associated with various measurements throughout the study.

Measurement	Accuracy	% Uncertainty	Measurement technique/ instrument
Load	± 0.1 kg	± 0.2	Electric eddy current type
Engine speed	± 10 rpm	± 0.1	Magnetic pickup type
Fuel flow	± 0.1 cu. cm	± 0.1	Volumetric measurement
Airflow	± 1 cu. cm	± 0.2	Orifice meter type
Crank angle	± 0.1 deg	± 0.01	Magnetic pickup type
Cylinder pressure	± 0.1 MPa	± 0.1	Piezoelectric pressure sensor
Temperature	± 1 °C	± 0.2	K-type thermocouple
Time	± 0.01 sec	± 0.2	Digital stopwatch
Carbon monoxide	± 0.01 % vol.	± 0.2	NDIR principle
Carbon dioxide	± 0.1 % vol.	± 0.15	NDIR principle
Hydrocarbon	± 5 ppm	± 0.2	NDIR principle
Nitrogen oxide	± 10 ppm	± 0.2	Electrochemical measurement

7.2.4 Assessment of potential environmental impacts

The potential environmental impacts (PEI) were determined through sustainability assessment using the life cycle assessment (LCA) methodology. An open-ended LCA methodology was followed for the analysis of the PEI for the functional unit under various boundary conditions as per ISO 14040/44 standard [4]. The emissions produced from a stationary diesel engine fueled with various blend fuels were designated as the unit for assessment in this study. A similar approach was implemented in a couple of recent studies [5, 6]. The functional unit, boundary conditions, and the various flows considered for the LCA-based assessment of PEIs are illustrated in Fig. 7.2. The flows of test fuels supplemented with air into the engine were considered as the input flows and the mass flows emissions produced from the engine fueled with various blend fuels were designated as the output flows for assessment in this study. An hour of engine operation at an invariant load condition was defined as the functional unit. The environmental impacts associated with the refining of diesel and kerosene and with the production, utilization of vegetable oil, and the rejection and collection of WVO were taken from inbuilt LCA databases in the software package. The environmental impacts of the fuel for different PEI categories are determined for two domains – one, the contribution from the production of the fuel, and second, the contribution from the combustion/ end-use of the fuel. Earlier studies have concluded that environmental impacts of biofuel are predominantly associated with emissions from the combustion, rather than the production steps [5]. The various impact categories are listed in Table 7.3. Accordingly, the LCA for the present study was performed in the GaBi® software package (Education version, Thinkstep, Germany) using the in-built US-LCA® and Eco-Invent® version 2.2 databases. A mid-point indicator-based CML 2001 baseline was used to determine the different PEI categories considered in the study.

Table 7.3 Different potential impact categories (PEI) considered for the sustainability assessment of various fuels.

Impact categories	Unit	Methodology
Global warning potential (GWP)	kg CO_2 equivalent	LCA
Acidification potential (AP)	kg SO_2 equivalent	LCA
Eutrophication potential (EP)	kg PO_4^- equivalent	LCA
Ozone depletion potential (ODP)	kg R-11 equivalent	LCA
Human toxicity potential (HTP)	kg DCB equivalent	LCA
Terrestrial eco-toxicity potential (TETP)	kg DCB equivalent	LCA
Total aquatic eco-toxicity potential (TATEP)	kg DCB equivalent	LCA
Abiotic depletion (Fossils) potential (ADP)	MJ	LCA

Note: R-11: Trichlorofluoromethane, DCB: 1.2-Dichlorobenzene

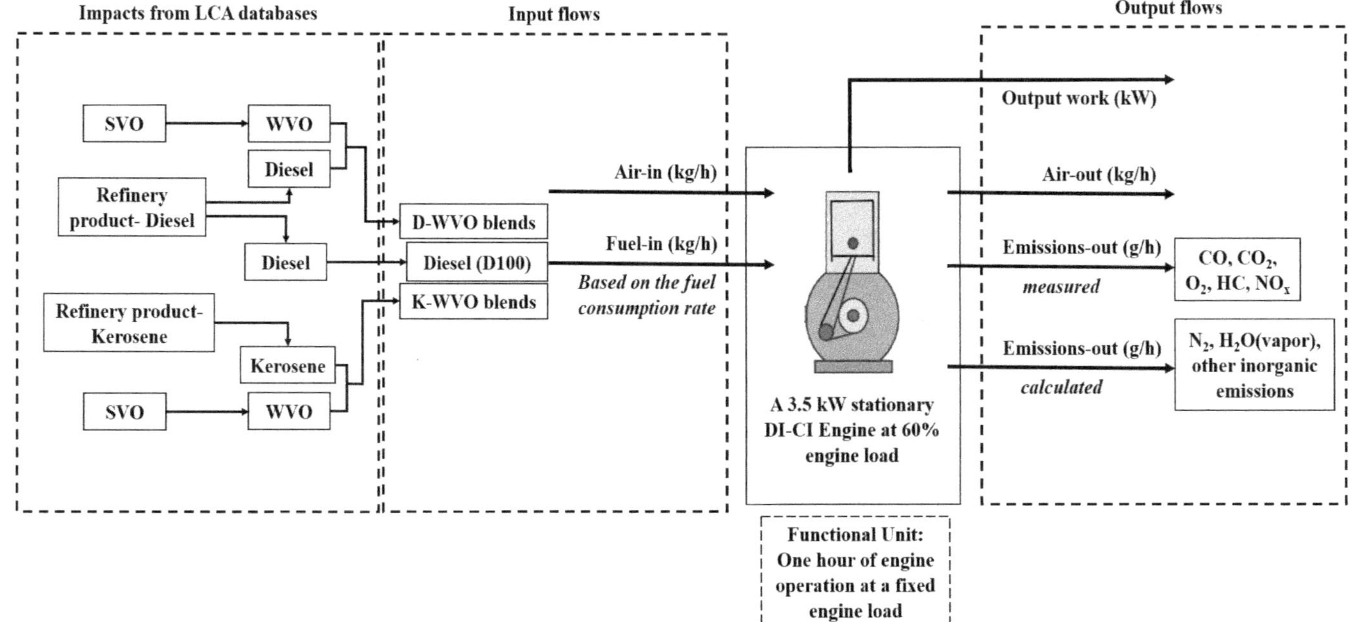

Fig. 7.2 The system boundary considered for combustion of **(a)** Diesel, **(b)** D-WVO50, and **(c)** K-WVO50 in a CI engine

7.3 Experimental findings

7.3.1 Comparison of fuel characteristics

The dissolution of WVO in both diesel and kerosene helped in preparing blends of the desired variations in WVO fractions. Various blends of WVO with diesel and kerosene were found to be quite stable with no phase separation occurring after 14 days. The six blends (3 D–WVO and 3 K–WVO) were analyzed for their fuel properties using ASTM standard test methods. Fig. 7.3 illustrates the comparison of the key fuel properties of D-WVO and K-WVO blends.

The density of the fuel determines the evaporation characteristics of fuel which affect its combustion and fuel economy. A denser fuel will cause higher mass flow for the same volume of fuel injected, thereby elevating the fuel consumption [7]. The denser WVO increased the densities of its blends with the increase in its fraction in the blend (Fig. 7.3(a)). However, the densities of K-WVO blends were very much lower than those for the diesel counterparts. The various K-WVO blends conformed to the IS 1460 standard for diesel fuel density range (0.820 - 0.845 g/cc), whereas, only D-WVO_0.20 and D-WVO_0.35 blends were within the standard limits. Thus, K-WVO blends were expected to show better fuel injection characteristics and lower fuel consumption than D-WVO blends. A minimum of 35 °C flashpoint is required to be considered for fueling a CI engine, as per IS 1460 standards. But, vegetable oil blends were known to have high flashpoints. Yet, they are reported available which claims that vegetable oil blends have satisfactorily powered a diesel engine [8]. Moreover, a higher flash point is desired for safer handling and transportation of fuel. The flashpoints of the blended fuels increased linearly with WVO addition (Fig. 7.3(b)). However, superior volatility of kerosene than diesel lowered the flashpoints of its blends than D-WVO blends.

Viscosity plays a pivotal role in the atomization and spray behaviors of a fuel. Higher viscosity causes poor atomization and minimizes jet penetration, but favors a reduction in fuel loss [9]. The viscosity of diesel and kerosene blends was seen to increase with the addition of WVO fractions (Fig. 7.3(a)). The standard range for diesel, as per IS 1460:2017, is 2.0-4.5 mm^2/s at 40 °C. Two K-WVO blends (K-WVO_0.20 and

K-WVO_0.35) and only one diesel blend (D-WVO_0.20) conformed to the standard range. So, better engine performance characteristics are expected of these particular blends. Calorific value is the measure of the energy obtainable from fuel from its combustion and determines the thermal efficiency of an engine. A fuel should possess a calorific value above 43,000 kJ/kg to be able to provide enough power to run an engine [1]. With the addition of WVO, the calorific value showed a steep decrease. However, the calorific values of all the WVO blends were above 43,000 kJ/kg (Fig. 7.3(b)). Thus, it can be confirmed that the blend fuels hold the ability to provide power to the CI engine similar, if not more, to neat diesel.

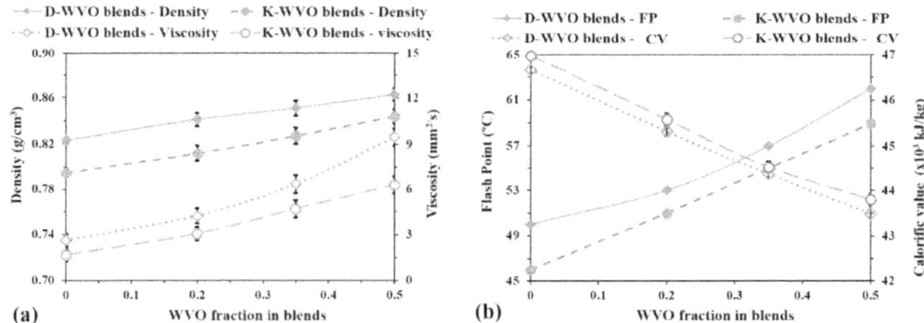

Fig. 7.3 Fuel properties comparison of D-WVO and K-WVO blends.

7.3.2 Comparison of engine combustion characteristics

The changes in the cylinder pressure for different WVO blends at 0.76 kW engine load conditions are presented in Fig. 7.4(a). For more clarity, 0.20 and 0.50 fractions containing WVO blends were selected for comparative analysis of combustion characteristics. The highest cylinder pressure (P_{max}) was recorded at around 387 °CA for the K-WVO_0.20 blend. The superior volatility of kerosene favored more atomization and caused increased burning of the K-WVO blends during the premixed combustion phase with higher resultant P_{max} output. The addition of WVO in diesel caused a slump in the P_{max} value along with a shift in the °CA. The profile of heat release for different test fuels is presented in Fig. 7.4(b). The combustion process was found to be more vigorous for the K-WVO blends as compared to D-WVO blends. The higher volatility of kerosene caused heightened combustion in K-WVO blends. The higher viscosity of WVO can impede complete atomization in D-WVO blends with increased

WVO fraction causing slower mixing of fuel and reducing the burned fraction of the fuel in the premixed phase of combustion [10]. As a result, the ignition delay was seen to increase with a rise in the WVO fraction in diesel blends, as shown in Fig. 7.4(c). As evident from Table 6.2, the cetane index was found to decrease with WVO addition in diesel. Thus, all of the above reasons may have collaboratively resulted in longer ignition delay periods for D-WVO blends than diesel. Also, it is known that burning lower fuel mass can cause longer ignition delay, slower evolution of pressure, and release of heat energy due to combustion [11]. However, in the case of K-WVO blends, the ignition delay improved with the increment in WVO in the blend. A comparatively higher cetane index of WVO than kerosene may be a prominent reason for this trend. Interestingly, almost similar ignition delay was observed for both diesel and kerosene blends containing 0.50 WVO fraction in them.

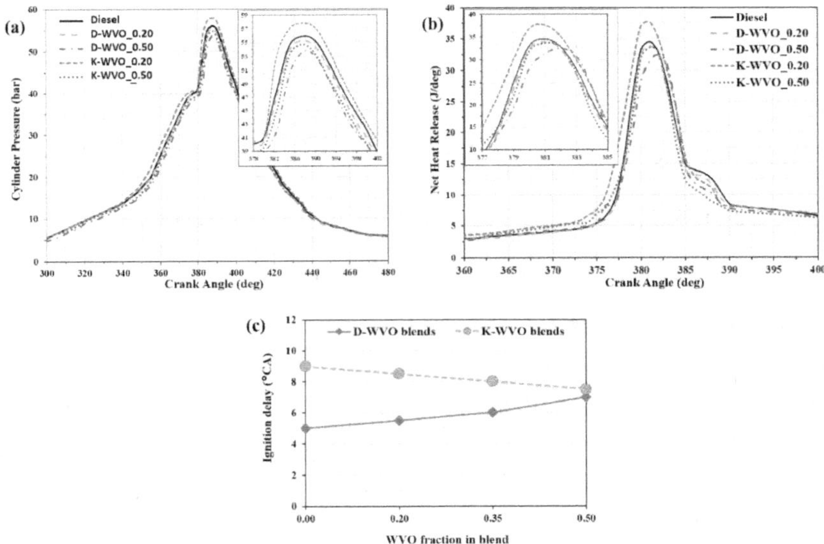

Fig. 7.4 Combustion characteristics – **(a)** cylinder pressure, **(b)** net heat release, and **(c)** ignition delay of D-WVO and K-WVO blends.

7.3.3 Comparison of engine performance characteristics

The engine performance was evaluated in terms of brake-specific fuel consumption (BSFC) and brake thermal efficiency (BTE) of the various blended fuels at the highest engine load condition considered in this study. BSFC is the rate at which the fuel is injected into the engine for producing the desired power at a specific engine load. An increase in the amount of fuel necessary to produce an output power increased the BSFC. BSFC of the various blends increased with the increment in WVO fraction in the blend, as presented in Fig. 7.5(a). This trend is due to the higher density and viscosity of most of the blend fuels than diesel. A higher density of fuel affects the fuel atomization and fuel-jet penetration inside the cylinder. Conversely, the inherent lubricity of vegetable oils is reported to reduce fuel loss. There exists a critical balance between the two phenomena and governs the rate of fuel consumption [9]. However, lower density and viscosity of the K-WVO_0.20 blend resulted in a 2.5% reduction in BSFC when compared to diesel. As compared to D-WVO blends, the K-WVO blends caused only a 3.3% increment in BSFC relative to diesel. Similar results are reported elsewhere [2, 3].

BTE denotes the efficiency of the engine to convert the chemical energy in the fuel into useful work upon combustion. It is dependent on the rate of fuel consumption and the calorific value of the fuel. The BTE of the blended fuel was found to drop with the rise in the WVO fraction in the blend. Lower calorific value and increased fuel consumption are the probable reasons. The K-WVO blends showed 3.3 - 4.8% higher BTE than neat diesel. Although the calorific values of these blends were lower than diesel, their densities and viscosities were very much similar to diesel. Also, owing to the reduction in fuel loss and superior volatility of kerosene caused faster air-fuel mixing, assisted the combustion process, and increased the BTE. The highest increase in BTE was recorded for the K-WVO_0.20 blend. Thus, it can be inferred that K-WVO blends demonstrated superior engine performance above their diesel counterparts.

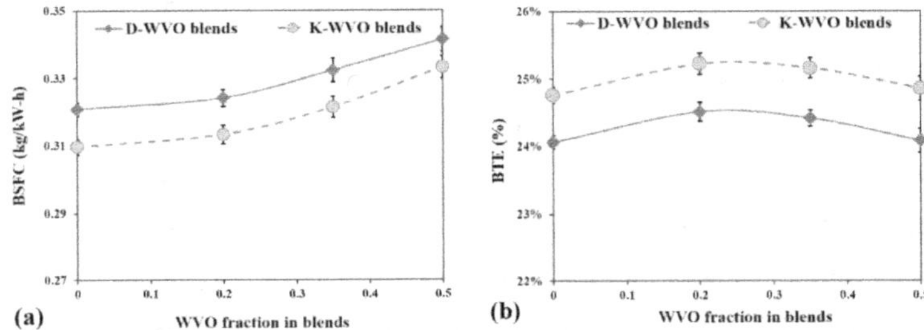

Fig. 7.5 Variations in the performance characteristics – **(a)** BSFC and **(b)** BTE of D-WVO blends and K-WVO blends at 3.0 kW brake power.

7.3.4 Comparison of engine emission characteristics

Brake-specific emissions were investigated for the various blended fuels in terms of carbon monoxide (CO), unburnt hydrocarbon (HC), and oxides of nitrogen (NO_x). The emission profile at the highest load condition considered in the study is presented and discussed in the subsequent paragraphs.

CO emission is the result of the incomplete combustion of fuel inside the cylinder. CO emission decreases with an increase in engine load. This is because a higher temperature inside the cylinder due to a rise in engine load improves the combustion efficiency of the engine [11]. It has been reported that CO emission reduces with the elevated oxygen content in the fuel. The reason is oxygenated fuel enhances the combustion of fuel and converts CO into CO_2 more efficiently [13]. Since WVO contains 10 - 12% oxygen, with the increment in the WVO fraction in the blended fuel, specific CO emission was found to decrease, as shown in Fig. 7.6(a). A rise in the oxygen mass fraction in the fuel blend with WVO addition assisted in the rapid oxidation of CO, resulting in a depletion of CO in the engine emission. In between the two types of blended fuels, a lower CO emission was recorded for the K-WVO blends than D-WVO blends. Relative to pure diesel, D-WVO blends caused a 13.4 - 18.1% CO reduction. Whereas for K-WVO blends the reduction in CO emission was between 26.7 - 30.7%, relative to diesel.

The HC emission is also produced due to the incomplete combustion of fuel. Literature suggests that more HC emission occurs at lower loads due to low in-cylinder temperatures than at high loads [11]. The specific HC emissions for D-WVO and K-WVO blends are presented in Fig. 7.6(b). With the addition of WVO fraction in diesel, specific HC emission increased, but it remained somewhat steady (11 - 12% increase) for the varying WVO fractions in the blend. However, HC emission showed an increasing trend with the increment in WVO fraction in the K-WVO blends. The probable reason would be higher volatility of kerosene and over-leaning effect at the power stroke which resulted in accumulation of fuel particles at the far corners of the engine cylinder. These fuel particles avoid complete combustion and are emitted as unburnt HC in the subsequent exhaust stroke. A similar explanation is present in the literature [3]. However, an HC reduction of 5.3 - 19.3% was achieved with K-WVO blends when to diesel fuel.

Higher NO_x emission takes place at high engine load conditions. It is known that NOx formation takes place at an elevated engine temperature and pressure, which occurs only at higher engine loads [11]. Fig. 7.6(c) presents the brake-specific NO_x emission at the highest engine load condition. It can be observed that for both D-WVO and K-WVO blends NO_x emission increased with the increase in WVO fraction. The increment in fuel-bound oxygen is the probable reason behind this trend. At higher in-cylinder temperatures enhanced oxidation of air-bound nitrogen takes place resulting in elevated NOx emissions [10]. The increment in the fuel-bound oxygen favored NOx formation. As a result higher WVO fractions caused more elevated NOx emission. Only the K-WVO_0.20 blend showed NO_x emission somewhat comparable to neat diesel. The diesel and kerosene blends containing 0.50 fraction WVO has caused a 9.1 - 11.6% rise in NO_x emission relative to diesel. A similar trend was reported elsewhere [9, 13].

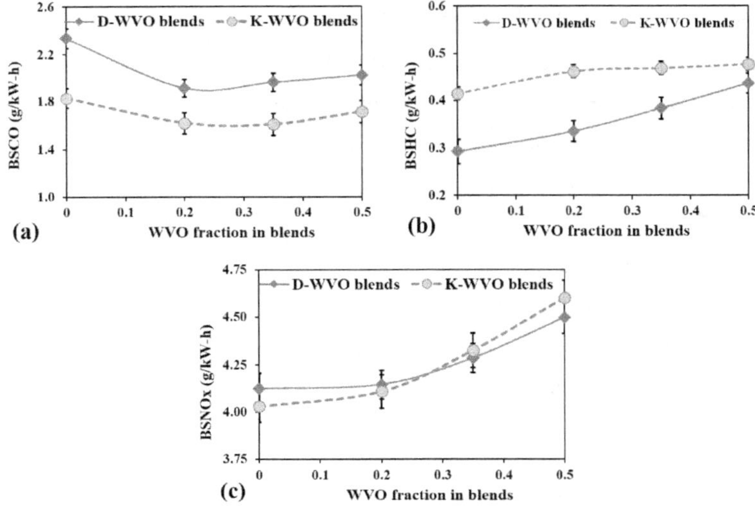

Fig. 7.6 Variations in the emission characteristics – **(a)** CO, **(b)** HC, and **(c)** NOx of D-WVO and K-WVO blends at 3.0 kW.

7.3.5 Comparison of engine running cost D-WVO and K-WVO blends

The WVO is known to have very little commercial value, therefore it holds the favorable economic reason for use as an alternative fraction in blend fuels [14]. The reported price of WVO is $ 0.19 per liter [15] compared to diesel at $ 0.89 per liter and kerosene oil at $ 0.49 a liter [16, 17]. The cost per hour of running a specific CI engine using blend fuel is also an important criterion to look into. Table 7.4 depicts the comparison of the running cost of the CI engine at the highest engine load used in the study. The product of the fuel cost (in $/L) with the rate of fuel consumption (L/h) at a fixed engine load was considered to calculate the engine running cost (in $/h). A 15.23 – 37.94% reduction in the engine running cost per hour was achievable for the three D-WVO blends. A 52.03 – 60.79% reduction in the engine running cost per hour can be achieved for 0.20 – 0.50 WVO fraction containing kerosene blends. Thus, it can be inferred that WVO addition in kerosene is linked with higher economic gains, than WVO addition in diesel.

Table 7.4 Comparison of the cost of the K-WVO blends against diesel and D-WVO blends.

Fuel and blends	Cost of D-WVO blends (per liter)[a,b]	Engine running cost for D-WVO blends[a,b,c]	% cost reduction for D-WVO blends[d]	Cost of K-WVO blends (per liter)[b]	Engine running cost for K-WVO blends[c]	% cost reduction for K-WVO blends[d]
D100	$ 0.891	$ 0.788	0.00 %	-----	-----	-----
K100	-----	-----	-----	$ 0.490	-----	-----
WVO	$ 0.195	-----	-----	$ 0.195	-----	-----
WVO_0.20	$ 0.752	$ 0.668	15.23 %	$ 0.431	$ 0.378	52.03 %
WVO_0.35	$ 0.647	$ 0.582	26.14 %	$ 0.387	$ 0.343	56.47 %
WVO_0.50	$ 0.543	$ 0.489	37.94 %	$ 0.343	$ 0.309	60.79 %

Note: D100: Diesel; K100: Kerosene; D-WVO: diesel-WVO blends; K-WVO: kerosene-WVO blends
[a] - Computed from values reported in online sources [15]
[b] - Price considered taken from Indian Oil Corporation Limited (2020) [16, 17]
[c] - Considered for one hour at the highest engine load setting
[d] - The cost reduction is relative to diesel fuel
The conversion rate considered: 1 Indian Rupee = 0.013 United States Dollar

7.3.6 Environmental impact assessment of D-WVO and K-WVO blends

The potential environmental impacts (PEIs) associated with various fuels are shown in Fig. 7.7. Being of renewable origin WVO is carbon neutral consequently WVO blends recorded lower global warming potential (GWP) compared to fossil fuels (Fig. 7.7(a)). The K-WVO_0.50 showed a 6.3% lower GWP than D-WVO_0.50. A probable reason can be the lower BSFC of K-WVO_0.50 than D-WVO_0.50. Due to no sulfur content of WVO compared to fossil fuels, reducing the acidification potential (AP) of the blends (Fig. 7.7(a)) [18]. The K-WVO_0.50 blend showed 4.7% lower AP than diesel, whereas D-WVO_0.50 showed only a 1.2% reduction. As evident from Fig. 7.7(b), all the three K-WVO blends showed quite similar EP. The WVO blends showed a reduction of 46.4 - 48.4% in ODP as compared to diesel fuel (Fig. 7.7(b)). Since the K-WVO blends contain a considerable renewable fraction, the ADP (fossil) reduction was between 21.8 − 47.8% as compared to diesel fuel. Diesel being a fossil fuel recorded the highest ADP of 36.6 MJ (Fig. 7.7(c)). K-WVO_0.50 recorded an ADP of 19.1 MJ, while D-WVO_0.50 recorded an ADP of 19.7 MJ for an hour of running the engine. The disparity between the two fossil fuels is directly linked with the lower fuel consumption of K-WVO_0.50 than D-WVO_0.50. The WVO blends showed around 22.9 - 47.4% reduction in HTP relative to diesel fuel (Fig. 7.7(c)). The two eco-toxicity parameters, TETP and AETP, also demonstrated similar trends (Fig. 7.7(d)).

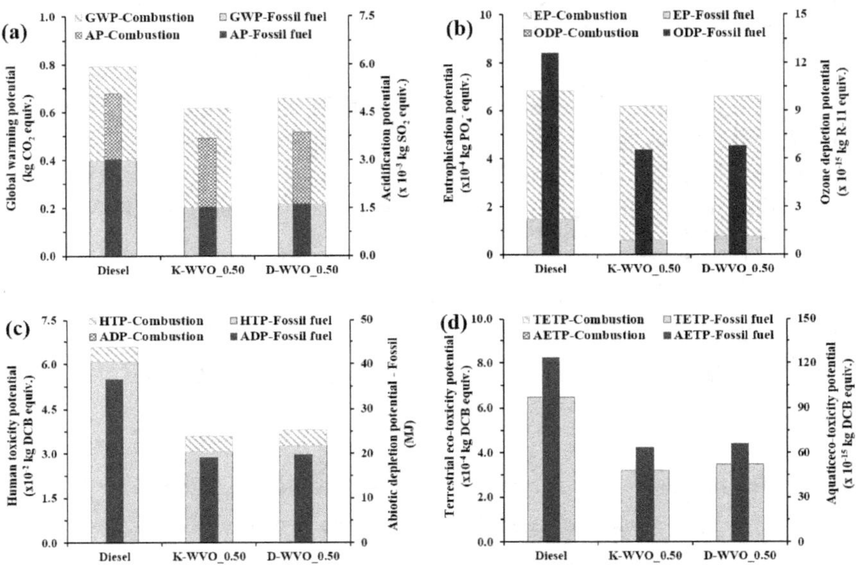

Fig. 7.7 Results of the sustainability assessment using LCA tool on one hour running of the compression ignition engine with the three fuels: diesel, D-WVO_0.50, and K-WVO_0.50.

Finally, to determine the most effective WVO blend fuel, a ranking and scoring matrix was utilized. The rank was varied between 1 to 7 for diesel, 3 D-WVO blends, and 3 K-WVO blends. The assessment parameters like P_{max}, NHR_{max}, and BTE were taken as higher the better, wherein the highest performing blend was given a rank of 7, while the lowest-performing one was marked as 1. Whereas, for parameters like BSFC, CO, HC, NO_x emissions, engine running cost, and PEIs were ranked as lower the better, wherein the blend with the lowest value was given a rank of 7, while the one with the highest value was marked as 1. Equal weightage was given to all the assessment parameters. The total score indicates the overall performance of a particular blend when equal weightage is given to all the assessment parameters. A high score signifies that the particular blend has performed better than most of the blends under investigation, while the reverse is true for a low score. The overall score of the matrix was determined by adding the ranks of each of the fuels, as presented in Table 7.5. From the matrix table, it can be seen that the K-WVO_0.20 blend got the highest score out of all the blends, followed by K-WVO_0.35 and K-WVO_0.50 blend. Thus, it can be inferred that the most suitable blend for substituting diesel in a CI engine is K-WVO_0.20.

Table 7.5 Rank and score of different assessment parameters.

Blend fuels	Rank for different assessment parameters									Total
	P_{max}	NHR_{max}	BSFC	BTE	CO	HC	NO_x	Engine running cost	PEIs	score
Diesel	5	6	5	4	1	7	6	1	1	36
D-WVO_0.20	3	4	3	3	2	6	3	2	2	28
D-WVO_0.35	2	2	2	2	4	5	2	3	4	26
D-WVO_0.50	1	1	1	1	6	4	1	4	6	25
K-WVO_0.20	7	7	7	7	3	3	7	5	3	49
K-WVO_0.35	6	5	6	6	5	2	5	6	5	46
K-WVO_0.50	4	3	4	5	7	1	4	7	7	42

The comprehensive and comparative assessment of WVO-based blends in a CI engine demonstrated that WVO addition in fossil fuels (diesel or kerosene) is an economic and a potentially viable strategy not only to valorize a common liquid waste – WVO but also to provide alternative fuel fraction at times when fossil fuel prices and availability are both unstable. WVO blends of diesel and kerosene are suitable to seamlessly operate in conventional diesel (CI) engines commonly used in power generation, transportation, and agricultural sector without modification.

The definition of a low-cost diesel substitute can bring beneficial impacts on the livelihood of the marginalized people below the poverty line in several developing and under-developed countries across the globe. In addition, the low-cost WVO-based diesel substitute blends are linked with lower potential environmental impacts relative to pure fossil fuels. Thus, the governmental agencies, looking for alternative fuels to replace fossil fuels, like diesel, have alternative fuel options fit for commercialization.

7.4 Summary

A common and abundant liquid waste was valorized as an alternate fuel fraction in this study. The effect of blending WVO with diesel and kerosene was investigated in this study. Three blends comprising of 0.20, 0.35, and 0.50 fraction WVO were separately formulated with diesel and kerosene. The fuel properties of such blends were determined and were found to be within the standards for diesel, except for the viscosity of the D-WVO_0.50 blend. With replacement in diesel with kerosene, the WVO blends demonstrated fuel properties close to hose for diesel. K-WVO_0.20 blend showed better engine performance, with lower BSFC and elevated BTE relative to diesel than the other blends. Lower density and viscosity, higher calorific value and superior volatility of kerosene is the probable reason. K-WVO blends resulted in lower CO, but higher HC emissions than diesel and D-WVO counterparts. The NO_x emission increased with the WVO addition in the blend. Only K-WVO_0.20 blend recorded NO_x emission similar to diesel. Blending WVO with fossil fuels has lowered the overall fuel cost. K-WVO blends can cause a 22 – 40% reduction in overall price as compared to similar D-WVO blends. Also, the PEIs associated with WVO blends were much lower compared to pure diesel. However, the K-WVO_0.50 blend showed maximized reduction in PEIs as compared to its diesel counterpart. After ranking and scoring the fuels on the different assessment parameters, K-WVO_0.20 was inferred as the most suitable diesel fuel replacement in a CI engine.

7.5 References

[1] Capuano, D., Costa, M., Di Fraia, S., Massarotti, N., Vanoli, L. (2017), "Direct use of waste vegetable oil in internal combustion engines", Renew Sustain Energy Rev, v. 69, pp. 759-770.

[2] Aydın H (2016), "Scrutinizing the combustion, performance and emissions of safflower biodiesel–kerosene fueled diesel engine used as power source for a generator", Energy Convers Manag, v. 117, pp. 400-409.

[3] Bayindir, H., Isık, M. K., Argunhan, Z., Yücel, H. L., Aydın, H. (2017), "Combustion, performance and emissions of a diesel power generator fueled with biodiesel-kerosene and biodiesel-kerosene-diesel blends", Energy, v. 123, pp. 241-251.

[4] Guinée, J. B. (2001), "Life cycle assessment: an operational guide to the ISO standards", (Final Ed.), Part 1 and 2, Ministry of Housing, Spatial Planning and Environment (VROM) and Centre of Environmental Science (CML), Den Haag and Leiden, Netherlands.

[5] Cavalcanti, E. J. C., Carvalho, M., Ochoa, A. A. V. (2019), "Exergoeconomic and exergoenvironmental comparison of diesel-biodiesel blends in a direct injection engine at variable loads", Energy Convers Manag, v. 183, pp. 450-461.

[6] Viornery-Portillo, E. A., Bravo-Diaz, B., Mena-Cervantes, V. Y. (2020), "Life cycle assessment and emission analysis of waste cooking oil biodiesel blend and fossil diesel used in a power generator", Fuel, v. 281, pp. 118739.

[7] Martin, M. L. J., Geo, V. E., Nagalingam, B. (2016), "Effect of fuel inlet temperature on cottonseed oil-diesel mixture composition and performance in a DI diesel engine", J Energy Institute, v. 90, pp. 563-573.

[8] Dabi, M., Saha, U. K. (2019), "Application potential of vegetable oils as alternative to diesel fuels in compression ignition engines: A review", J Energy Institute, v. 92, pp. 1710-1726.

[9] Agarwal, A. K., Dhar, A. (2013), "Experimental investigations of performance, emission and combustion characteristics and combustion characteristics of Karanja oil blends fueled DICI engine", Renew Energy, v. 52, pp. 283–91.

[10] Prabu, S. S., Asokan, M. A., Roy, R., Francis, S., Sreelekh, M. K, (2017), "Performance, Combustion and Emission Characteristics of Diesel Engine fueled with Waste Cooking Oil Bio-diesel or diesel blends with Additives", Energy, v. 122, pp. 638-648.

[11] Yesilyurt, M. K. (2019), "The effects of the fuel injection pressure on the

performance and emission characteristics of a diesel engine fuelled with waste cooking oil biodiesel-diesel blends", Renew Energy, v. 132, pp. 649–666.

[12] Mat, S. C., Idroas, M. Y., Hamid, M. F., Zainal, Z. A. (2018), Performance and emissions of straight vegetable oils and its blends as a fuel in diesel engine: A review", Renew. Sustain. Energy Rev., v. 82 pp. 808-823.

[13] Huang, H., Teng, W., Liu, Q., Zhou, C., Wang, Q., Wang, X. (2016), "Combustion, performance and emission characteristics of a diesel engine under low-temperature combustion of pine oil–diesel blends", Energy Conver Manag, v. 128, pp. 317-26.

[14] Majhi, S., Ray, S. (2016), "A study on production of biodiesel using a novel solid oxide catalyst derived from waste", Environ Sci Pollut Res, v. 23, pp. 9251-9259.

[15] Akshatha, M. (2018), "Hotels unaware that used cooking oil can yield biodiesel", Economic Times, https://economictimes.indiatimes.com/industry/energy/oil-gas/hotels-unaware-that-used-cookingoil-can-yield-biodiesel/articleshow/62418971.cms. Accessed 18 November 2018.

[16] Indian Oil Corporation Limited. 2020. Diesel historic price. January 2020. Accessed June 16, 2020.

[17] Indian Oil Corporation Limited. 2020. Kerosene historic price. January 2020. Accessed June 16, 2020. https://www.iocl.com/Product_PreviousPrice/KerosenSubsidyPreviousPrice.aspx.

[18] Hossain, A. K., Davies, P. A. (2010), "Plant oils as fuels for compression ignition engines: A technical review and life-cycle analysis", Renew Energy, v.35, pp. 1-13.

Chapter 8

SUMMARY AND CONCLUSIONS

8.1 Summary of the Thesis

In recent years, depleting fossil fuel reserves, increasing fuel demand, and instability in fuel supply and prices are the burning issues that the energy sector is facing. One effort to solve these problems is to replace existing fossil fuels with renewable and alternative fuels. Research on replacing diesel has been ongoing for many decades and various substitutes, in terms of blend fractions, have been proposed. In this context, vegetable oils and their derivatives (biodiesels) are the most sought-after ones. However, such fuels are often associated with food versus fuel disputes and higher overall fuel costs. Thus, utilizing a waste-derived fuel can provide a sustainable and cost-effective solution to this problem.

Waste vegetable oil (WVO), if and when managed wisely, can become a potential feedstock for alternative fuels. WVO is most often valorized as WVO-biodiesel (WVO-BD) and blended with diesel to fuel compression ignition (CI) engines. However, WVO-BD can be easily blended with other fossil fuels too. Thus, an optimization study of this sort, as presented in Chapter 3, showed that higher WVO-BD fractions should be mixed with kerosene-grade fossil fuel to achieve density and calorific values similar to diesel. The model-optimized blend presented better engine performance characteristics with comparable engine emissions were shown relative to diesel.

In Chapter 4, a comparative study between diesel-WVO (D-WVO) and diesel-WVO-BD (D-WVO-BD) blends exhibited very little difference in terms of performance and emission characteristics in a 3.7 kW stationary compression ignition (CI) engine. WVO addition to diesel drastically reduced the engine running cost (up to 36%), relative to diesel, when compared to D-WVO-BD blends (with an 18.5% reduction). Thus, the utilization of chemically unmodified WVO as an alternative fraction in diesel was achieved and proposed as a sustainable and economic strategy to fuel small CI engines to replace diesel.

After D-WVO blends were proposed as an economic diesel fuel substitute in light-duty CI engines, WVO, of varying thermal exposure time, was optimized to be used as a diesel blend fraction in Chapter 5. To optimize WVO fractions in diesel for lowering the fuel consumption (FC) and improving the thermal efficiency (TE) of a CI engine using a response surface approach was the objective. A diesel blend containing 0.50 fraction 48 hours thermally exposed WVO at 14.72 Nm engine torque was found to have lower FC (0.5092 kg/h) and higher TE (23.84%) among other factor-level settings. The experimentally obtained FC and TE were very close to the model-predicted values. Also, at a fixed engine torque (or load) the optimized FC and TE for D-WVO blends were found to be lower than various diesel-vegetable oil blends presented in recent literature.

In Chapter 6, a full-scale study on D-WVO blends (in which WVO with varying thermal exposure time was blended with diesel at different fractions) to fuel a CI engine is presented. A general reduction in TE and increase in FC was found for D-WVO blends, compared to diesel. Lower carbon monoxide (CO) and comparable oxides of nitrogen (NO_x) emissions were recorded for the D-WVO blends relative to diesel. In the same chapter, WVO was further mixed with kerosene (by replacing diesel) at varying WVO fractions and the engine performance and emission studies for kerosene-WVO (K-WVO) blends were carried out. An improvement in the engine performance, with increased TE and lowered FC, was achieved for the K-WVO blends, compared to diesel. WVO blends with both diesel and kerosene were suitable for fueling a small CI engine with no operational difficulties.

Finally, in Chapter 7, a comparative study between D-WVO and K-WVO blends was carried out to determine the most suitable fuel for blending WVO and replacing diesel in a small CI engine. Both engine combustion and performance were better for K-WVO blends, while better HC emissions were achieved for D-WVO blends. CO and NO_x emissions were very much comparable for similar WVO fractions in the blends. K-WVO blends were more economical, with up to 69% reduction in the engine running cost is possible, relative to diesel. The D-WVO and K-WVO blends were linked with lesser potential environmental impacts, compared to diesel. In between blends, the K-WVO_0.50 blend has comparatively lesser potential environmental impacts than the D-WVO_0.50 blend.

8.2 Summary of contributions

The main contributions of the thesis are summarized below:

- A waste vegetable oil-biodiesel (WVO-BD)-based alternative blend fuel is defined, modeled, and optimized to substitute diesel for fueling an unmodified compression ignition engine.

- A comparative analysis of WVO-BD and WVO blends with diesel has shown comparable engine performance and emissions, while the diesel-WVO blends significantly reduced engine running costs compared to diesel.

- The direct blending of chemically unmodified WVO with diesel to operate a CI engine is a key highlight of the thesis.

- The fuel consumption and thermal efficiency of the WVO-diesel blend are modeled and optimized to provide lower fuel consumption and higher thermal efficiency in a compression ignition engine.

- WVO is also deliberated as an alternative fuel fraction in kerosene, which diminished the disparity between WVO blends fuel and diesel.

- The kerosene-WVO (K-WVO) blends are superior diesel fuel substitutes than diesel-WVO blends.

- The assessment of potential environmental impacts shows WVO blends are environmentally benign than diesel fuel.

Additionally, the various studies on WVO blends (as presented in the thesis) are conclusive in promoting WVO as a suitable vegetable oil replacement in fossil fuel blends for fueling small CI engines. The WVO blends (especially K-WVO blends) have shown potential in replacing diesel for operating light-duty CI engines, those utilized in agriculture, and standalone power generators. These blends can easily substitute 100% diesel in low-duty engines while being more economical and environmentally benign than diesel.

8.3 Future scopes in the Thesis

Building upon this thesis, the following issues may be considered as future scopes in this area of research:

- Engine wear characteristics due to WVO blends in detail, with emphasis on deposits inside the engine cylinder. Also, tribological studies on the changes in the injector nozzle, engine valves, and fuel lines.

- The role of various synthetic additives in WVO blends in engine combustion, performance, and emission characteristics is another scope of research.

LIST OF PUBLICATIONS

Following are the publications derived out of the undertaken Ph.D. work:

International Journals:

1. Pritam Dey, Srimanta Ray, "Valorization of waste vegetable oil (WVO) for utilization as diesel blends in CI engine – performance and emission studies", Energy Sources, Part A: Recovery, Utilization, and Environmental Effects, 1-14 (2019). (SCI indexed)

2. Pritam Dey, Srimanta Ray, "Comparative analysis of waste vegetable oil versus transesterified waste vegetable oil in diesel blend as alternative fuels for compression ignition engine" Clean Technologies and Environmental Policy, 22(7), 1517-1530 (2020). (SCI-E indexed)

3. Pritam Dey, Srimanta Ray, "Optimization of Waste Vegetable Oil–Diesel Blends for Engine Performance: A Response Surface Approach", Arabian Journal of Science and Engineering, 45, 7725-7739 (2020). (SCI-E indexed)

4. Pritam Dey, Srimanta Ray, Abhishek Newar, "Defining a waste vegetable oil-biodiesel based diesel substitute blend fuel by response surface optimization of density and calorific value", Fuel 283, 118978 (2021). (SCI indexed)

5. Pritam Dey, Srimanta Ray, "Comprehensive assessment of sustainable low-cost waste vegetable oil-based blend as a diesel substitute", Clean Technologies and Environmental Policy, 23, 1521-1536 (2021). (SCI-E indexed)

6. Pritam Dey, Srimanta Ray, "Valorization of waste vegetable oil-based blend fuels as sustainable diesel replacement: Comparison between blending with diesel versus kerosene", Clean Technologies and Environmental Policy, 1-14 (2021). (SCI-E indexed)

Presented in International Conferences/ Symposiums:

1. Pritam Dey, Srimanta Ray, "Valorization of waste vegetable oil for utilization as diesel blends in CI engine – performance and emission studies", Proceedings of the 1st International Conference on Advances and Challenges for Sustainable Ecosystem 2018 (ICACSE 2018), 6-8 December, 2018.

2. Pritam Dey, Srimanta Ray, Gopal Mugeraya, "Performance and emission studies of waste vegetable oil-kerosene blends in CI engine – A sustainable alternative to diesel", Proceedings of the 11th International Exergy, Energy, and Environment Symposium (IEEES'11), 14-18 July, 2019.

Presented in National Conferences:

1. Pritam Dey, Srimanta Ray, "Evaluation of Performance and Fuel Economy of Various Diesel Blends in a Compression Ignition Engine", Proceedings of the National Conference on Sustainable Infrastructure Development and Management (SIDM 2019), 22-23 February, 2019.

2. Abhishek Newar, Pritam Dey, Srimanta Ray, "Comparison of Physical and Emission Properties of Blends of Waste Vegetable Oil and Waste Vegetable oil Biodiesel with Diesel", Proceedings of the National Conference on Sustainable Infrastructure Development and Management (SIDM 2019), 22-23 February, 2019.

APPENDIX-B

ABOUT THE SCHOLAR

Mr. Pritam Dey, a resident of Agartala, Tripura, completed his (10+2) schooling from Kendriya Vidyalaya No. 2, ONGC Colony, Agartala, Tripura in 2007. He pursued Bachelor of Technology in Bitechnology from SRM Institute of Science and Technology, SRM University, Tamil Nadu and completed the course in First Class with distinction in 2011. Upon qualifyling GATE-2011, he pursued Master of Technology in Chemical Engineering from Malviya National Institute of Technology Jaipur, Rajasthan. He was awarded the Gold Medal during his Masters graduation on 10th July 2013. He joined National Institute of Technolgy Agartala, Tripura as a full-time research scholar in the Chemical Engineering Department in 2017, under the able supervision of Dr. Srimanta Ray, Assistant Professor and Head, Chemical Engineering Department.

The research goal selected by Mr. Dey was in the area of waste mitigation and valorization. During his time in NIT Agartala, he has presented his works in several national and international conferences and symposia. He was awarded prize for the Best Oral Presentation at the International Conference on Advances and Challenges for Sustainable Ecosystem 2018 held at National Institute of Technology Tiruchirapalli, Tamil Nadu between 6-8 December, 2018. He was also awarded the first prize for Poster Presentation at the Engineer's Day Celebration, organized by The Institution of Engineers (India) – Tripura State Chapter on 15th September, 2018.

His research interests are in the areas of wastewater remediation, biofuels, catalysis, pyrolysis, and alternative liquid fuels.

VALORIZATION OF LIQUID WASTE AS ALTERNATIVE FUEL

Thesis submitted to National Institute of Technology Agartala
for the award of the degree of

Doctor of Philosophy

by

Pritam Dey

(Enrollment No: 16EDCHR004)

Under the Supervision of

Dr. Srimanta Ray

Assistant Professor

Department of Chemical Engineering, NIT Agartala

and

Prof. (Dr.) Gopal Mugeraya

Professor

Department of Chemical Engineering, NITK Surathkal

DEPARTMENT OF CHEMICAL ENGINEERING

NATIONAL INSTITUTE OF TECNOLOGY AGARTALA

January 2022

i

Chapter 8

SUMMARY AND CONCLUSIONS

8.1 Summary of the Thesis

In recent years, depleting fossil fuel reserves, increasing fuel demand, and instability in fuel supply and prices are the burning issues that the energy sector is facing. One effort to solve these problems is to replace existing fossil fuels with renewable and alternative fuels. Research on replacing diesel has been ongoing for many decades and various substitutes, in terms of blend fractions, have been proposed. In this context, vegetable oils and their derivatives (biodiesels) are the most sought-after ones. However, such fuels are often associated with food versus fuel disputes and higher overall fuel costs. Thus, utilizing a waste-derived fuel can provide a sustainable and cost-effective solution to this problem.

Waste vegetable oil (WVO), if and when managed wisely, can become a potential feedstock for alternative fuels. WVO is most often valorized as WVO-biodiesel (WVO-BD) and blended with diesel to fuel compression ignition (CI) engines. However, WVO-BD can be easily blended with other fossil fuels too. Thus, an optimization study of this sort, as presented in Chapter 3, showed that higher WVO-BD fractions should be mixed with kerosene-grade fossil fuel to achieve density and calorific values similar to diesel. The model-optimized blend presented better engine performance characteristics with comparable engine emissions were shown relative to diesel.

In Chapter 4, a comparative study between diesel-WVO (D-WVO) and diesel-WVO-BD (D-WVO-BD) blends exhibited very little difference in terms of performance and emission characteristics in a 3.7 kW stationary compression ignition (CI) engine. WVO addition to diesel drastically reduced the engine running cost (up to 36%), relative to diesel, when compared to D-WVO-BD blends (with an 18.5% reduction). Thus, the utilization of chemically unmodified WVO as an alternative fraction in diesel was achieved and proposed as a sustainable and economic strategy to fuel small CI engines to replace diesel.

After D-WVO blends were proposed as an economic diesel fuel substitute in light-duty CI engines, WVO, of varying thermal exposure time, was optimized to be used as a diesel blend fraction in Chapter 5. To optimize WVO fractions in diesel for lowering the fuel consumption (FC) and improving the thermal efficiency (TE) of a CI engine using a response surface approach was the objective. A diesel blend containing 0.50 fraction 48 hours thermally exposed WVO at 14.72 Nm engine torque was found to have lower FC (0.5092 kg/h) and higher TE (23.84%) among other factor-level settings. The experimentally obtained FC and TE were very close to the model-predicted values. Also, at a fixed engine torque (or load) the optimized FC and TE for D-WVO blends were found to be lower than various diesel-vegetable oil blends presented in recent literature.

In Chapter 6, a full-scale study on D-WVO blends (in which WVO with varying thermal exposure time was blended with diesel at different fractions) to fuel a CI engine is presented. A general reduction in TE and increase in FC was found for D-WVO blends, compared to diesel. Lower carbon monoxide (CO) and comparable oxides of nitrogen (NO_x) emissions were recorded for the D-WVO blends relative to diesel. In the same chapter, WVO was further mixed with kerosene (by replacing diesel) at varying WVO fractions and the engine performance and emission studies for kerosene-WVO (K-WVO) blends were carried out. An improvement in the engine performance, with increased TE and lowered FC, was achieved for the K-WVO blends, compared to diesel. WVO blends with both diesel and kerosene were suitable for fueling a small CI engine with no operational difficulties.

Finally, in Chapter 7, a comparative study between D-WVO and K-WVO blends was carried out to determine the most suitable fuel for blending WVO and replacing diesel in a small CI engine. Both engine combustion and performance were better for K-WVO blends, while better HC emissions were achieved for D-WVO blends. CO and NO_x emissions were very much comparable for similar WVO fractions in the blends. K-WVO blends were more economical, with up to 69% reduction in the engine running cost is possible, relative to diesel. The D-WVO and K-WVO blends were linked with lesser potential environmental impacts, compared to diesel. In between blends, the K-WVO_0.50 blend has comparatively lesser potential environmental impacts than the D-WVO_0.50 blend.

8.2 Summary of contributions

The main contributions of the thesis are summarized below:

- A waste vegetable oil-biodiesel (WVO-BD)-based alternative blend fuel is defined, modeled, and optimized to substitute diesel for fueling an unmodified compression ignition engine.

- A comparative analysis of WVO-BD and WVO blends with diesel has shown comparable engine performance and emissions, while the diesel-WVO blends significantly reduced engine running costs compared to diesel.

- The direct blending of chemically unmodified WVO with diesel to operate a CI engine is a key highlight of the thesis.

- The fuel consumption and thermal efficiency of the WVO-diesel blend are modeled and optimized to provide lower fuel consumption and higher thermal efficiency in a compression ignition engine.

- WVO is also deliberated as an alternative fuel fraction in kerosene, which diminished the disparity between WVO blends fuel and diesel.

- The kerosene-WVO (K-WVO) blends are superior diesel fuel substitutes than diesel-WVO blends.

- The assessment of potential environmental impacts shows WVO blends are environmentally benign than diesel fuel.

Additionally, the various studies on WVO blends (as presented in the thesis) are conclusive in promoting WVO as a suitable vegetable oil replacement in fossil fuel blends for fueling small CI engines. The WVO blends (especially K-WVO blends) have shown potential in replacing diesel for operating light-duty CI engines, those utilized in agriculture, and standalone power generators. These blends can easily substitute 100% diesel in low-duty engines while being more economical and environmentally benign than diesel.

8.3 Future scopes in the Thesis

Building upon this thesis, the following issues may be considered as future scopes in this area of research:

- Engine wear characteristics due to WVO blends in detail, with emphasis on deposits inside the engine cylinder. Also, tribological studies on the changes in the injector nozzle, engine valves, and fuel lines.

- The role of various synthetic additives in WVO blends in engine combustion, performance, and emission characteristics is another scope of research.